Architectural glass to resist seismic and extreme climatic events

Related titles:

Strengthening and rehabilitation of civil infrastructures using fibre-reinforced polymer (FRP) composites
(ISBN 978-1-84569-448-7)
The book discusses the mechanical and in-service properties, the relevant manufacturing techniques and aspects related to externally bonded FRP composites to strengthen/rehabilitate/retrofit civil engineering structural materials. The book focuses on: Mechanical properties of the FRP materials used; Analysis and design of strengthening/rehabilitating/retrofitting beams and columns manufactured from reinforced concrete (RC), metallic and masonry materials; Failure modes of strengthening systems; Site preparation of the two adherend materials; Durability issues; Quality control, maintenance and repair of structural systems; Some case studies.

Durability of concrete and cement composites
(ISBN 978-1-85573-940-6)
Concrete and other cement-based composites are by far the most widely used man-made construction materials in the world. However, major problems of infrastructure deterioration have been caused by unanticipated premature degradation of these materials. This book provides a recent review of several of the main forms of degradation, examining what is known about their causes and control. Trends in modelling and prediction of service lives are also examined.

Inspection and monitoring techniques for bridges and civil structures
(ISBN 978-1-85573-939-0)
The safety, maintenance and repair of bridges and buildings depend on effective inspection and monitoring techniques. These methods need to be able to identify problems, often hidden within structures, before they become serious. This important collection reviews key techniques and their applications to bridges, buildings and other civil structures. Chapters review ways of testing corrosion in concrete components; ways of testing wood components within civil structures; and acoustic techniques and their use to assess bridges in particular.

Details of these and other Woodhead Publishing materials books can be obtained by:

- visiting our web site at www.woodheadpublishing.com
- contacting Customer Services (e-mail: sales@woodheadpublishing.com; fax: +44 (0) 1223 893694; tel: +44 (0) 1223 891358 ext. 130; address: Woodhead Publishing Limited, Abington Hall, Granta Park, Great Abington, Cambridge CB21 6AH, UK)

If you would like to receive information on forthcoming titles, please send your address details to: Francis Dodds (address, tel. and fax as above; e-mail: francis.dodds@woodheadpublishing.com). Please confirm which subject areas you are interested in.

Architectural glass to resist seismic and extreme climatic events

Edited by
Richard A. Behr

CRC Press
Boca Raton   Boston   New York   Washington, DC

WOODHEAD PUBLISHING LIMITED
Oxford   Cambridge   New Delhi

Published by Woodhead Publishing Limited, Abington Hall, Granta Park, Great Abington, Cambridge CB21 6AH, UK
www.woodheadpublishing.com

Woodhead Publishing India Private Limited, G-2, Vardaan House, 7/28 Ansari Road, Daryaganj, New Delhi – 110002, India

Published in North America by CRC Press LLC, 6000 Broken Sound Parkway, NW, Suite 300, Boca Raton, FL 33487, USA

First published 2009, Woodhead Publishing Limited and CRC Press LLC
© 2009, Woodhead Publishing Limited
The authors have asserted their moral rights.

British Library Cataloguing in Publication Data
A catalogue record for this book is available from the British Library.

Library of Congress Cataloging in Publication Data
A catalog record for this book is available from the Library of Congress.

Woodhead Publishing ISBN 978-1-84569-369-5 (book)
Woodhead Publishing ISBN 978-1-84569-685-6 (e-book)
CRC Press ISBN 978-1-4398-0170-3
CRC Press order number: N10037

The publishers' policy is to use permanent paper from mills that operate a sustainable forestry policy, and which has been manufactured from pulp which is processed using acid-free and elemental chlorine-free practices. Furthermore, the publishers ensure that the text paper and cover board used have met acceptable environmental accreditation standards.

Typeset by Data Standards Ltd, Frome, Somerset, UK
Printed by TJ International Limited, Padstow, Cornwall, UK

Contents

Contributor contact details

(* = main contact)

Editor

Dr Richard A. Behr, PE
Department of Architectural
Engineering
The Pennsylvania State University
216 Engineering Unit A
University Park
PA 16802
USA
Email: behr@engr.psu.edu

Chapter 1

Robert E. Bachman, SE*
R. E. Bachman Consulting
Structural Engineers
25152 La Estrada Drive
Laguna Niguel
California 92677
USA
Email: REBachmanSE@aol.com

Susan M. Dowty, SE
S. K. Ghosh Associates Inc.
43 Vantis Drive
Aliso Viejo
California 92656
USA

Chapter 2

Dr Ali M. Memari, PE*
Department of Architectural
Engineering
The Pennsylvania State University

104 Engineering Unit A
University Park
PA 16802
USA
Email: memari@engr.psu.edu

Thomas A. Schwartz, PE
President, Head of Building
Technology
Simpson Gumpertz & Heger Inc.
41 Seyon Street
Suite 500
Waltham
MA 02453
USA
Email: taschwartz@sgh.com

Chapter 3

Richard S. Flood, AIA/CSI
Matrix IMA
98 Battery Street
Suite 500
San Francisco
California 94111
USA
Email: rick@matrix-ima.com

Chapter 4

R. H. Davies, PE*
Simpson Gumpertz & Heger Inc.
19 W 34th Street
Suite 1000
New York
NY 10001
USA
Email: rhdavies@sgh.com

Niklas W. Vigener, PE
Principal
Simpson Gumpertz & Heger Inc.
Park Plaza 1
2101 Gaither Road
Suite 250
Rockville
MD 20850
USA
Email: nwvigener@sgh.com

Chapter 5

Dr Kishor C. Mehta, PE
P. W. Horn Professor of Civil
Engineering
Texas Tech University
Lubbock
Texas 79409-1023
USA
Email: Kishor.Mehta@ttu.edu

Chapter 6

Chris Barry
Director, Technical Services
Pilkington NA Inc.
811 Madison Avenue
Toledo
OH 43604-5684
USA
Email: christopher.barry@
us.pilkington.com

Chapter 7

David B. Hattis
Building Technology Inc.
1109 Spring Street
Silver Spring, MD 20910
USA
Email: dbhattis@bldgtechnology.
com

Chapter 8

Dr Joseph E. Minor, PE
Consulting Engineer
PO Box 603
Rockport
TX 78381
USA
Email: josephminor@sbcglobal.net

Chapter 9

Scott A. Warner
Executive Vice President
Architectural Testing Inc.
130 Derry Court
York, PA 17406-8405
USA
Email: swarner@archtest.com

Preface

Architectural glass components in building envelope systems are commonly classified as 'nonstructural' or 'architectural' components. Such a classification implies that they are non-load-bearing, which is certainly not the case in reality. When a building is challenged by natural hazards such as earthquakes, heavy snowstorms, and severe windstorms, architectural glass components (i.e. windows, spandrel panels, glazed doors, overhead glazing, skylights, etc.) are often at the very frontline of that building's structural defenses. For this important reason architectural glass components must be designed as structural building components. Classifying architectural glass components as 'nonstructural' often leaves them in a gap of design attention between the architect and the structural engineer. This gap needs to be closed.

The purpose of this book is to provide the building envelope designer with a comprehensive resource document and design guide that will enable him/her to design architectural glass components to resist seismic, snow, and wind effects in accordance with the latest model building code provisions and industry standards in the United States. The scope of this book is limited to the naturally occurring phenomena of earthquakes, snowstorms, and windstorms. Man-made hazards that could necessitate blast design and/or ballistic design of architectural glass were not included in this book because they are special design scenarios requiring special treatments.

This book is intended to be used as a comprehensive resource document and design guide for the structural design of architectural glass by architects and engineers responsible for building envelope performance design. Numerous completely worked example problems and illustrative figures are included throughout to make this a user-friendly design guide and resource document. Also included are some state-of-the-art research results that are now pushing the envelope of architectural glass design practice. Such information in chapter bodies and reference lists at the end of each chapter provides the reader with additional paths for further investigation. Each chapter was intended to be as self-contained as possible, but this could not always be accomplished entirely within reasonable chapter length limitations. Therefore, to be totally thorough, it is recommended that those using this book for building envelope design projects also have available the latest editions of the following four documents: (1) ASCE 7, *Minimum Design Loads for Buildings and Other Structures*; (2) the International Building Code; (3) ASTM E1300, *Standard Practice for Determining Load Resistance of Glass in Buildings*; and (4) AAMA 501.05, *Methods of Tests for Exterior Walls*. For most common situations, however, a designer could complete the required project work with just a copy of this book and a copy of ASCE 7 on hand.

I wish to express my sincere gratitude to all of the contributing authors for their diligent efforts in working with me and the staff at Woodhead Publishing to complete their strong chapter contributions to this book. All of the chapter authors are leading authorities in their fields, and I am most appreciative to have had the opportunity to work with them to complete this project in style.

Richard A. Behr, PhD, PE
University Park, PA

1

Building code seismic requirements for architectural glass: the United States

R. E. BACHMAN, R. E. Bachman Consulting Structural Engineers, USA and S. M. DOWTY, S. K. Ghosh Associates Inc., USA

Abstract: This chapter describes the current building code seismic requirements for architectural glass in the United States. The chapter first reviews the development of seismic requirements for nonstructural components in the United States. It then discusses the specific requirements for architectural glass which are treated as a subset of nonstructural components. These requirements primarily focus on providing an adequate clearance gap around the edge of the glass that would accommodate the anticipated horizontal relative displacements of a building during design earthquake events.

Key words: building codes, seismic requirements, nonstructural components, architectural glass.

1.1 Introduction

The most current building code enforced in most jurisdictions in the United States is the *2006 International Building Code* (IBC, 2006). The 2006 IBC references the 2005 edition of the standard *Minimum Design Loads for Buildings and Other Structures* prepared by the American Society of Civil Engineers (ASCE, 2005) for its seismic provisions. ASCE 7-05 contains specific requirements for nonstructural components including requirements for architectural glass. There have been many significant examples of poor performance of architectural glass in past earthquakes (see Figs 1.1(a) and (b)). These have resulted in the development of new seismic standards and code requirements for architectural glass in the United States. This chapter provides the background on the development of US building code seismic requirements in Section 1.2 and an overview of what the building seismic requirements are in Section 1.3. In Section 1.4, the seismic requirements for nonstructural components are discussed and in Section 1.5, the specific

(a)

(b)

1.1 (a) Broken store windows after 1 October 1987 Whittier Narrows Earthquake (photos taken by Susan Dowty). (b) Broken curtain wall glazing after 2001 Nisqually Earthquake (photo taken by Tom Reese, *Seattle Times*, March 2001).

requirements for architectural glass are described. Future trends in seismic requirements for architectural glass including performance-based design is described in Section 1.6, other sources of information are described in Section 1.7, and references are provided in Section 1.8.

1.2   Background

The most current building code enforced in most jurisdictions in the United States is the *2006 International Building Code* (IBC, 2006). The 2006 IBC references the 2005 edition of the standard *Minimum Design Loads for Buildings and Other Structures* prepared by the American Society of Civil Engineers (ASCE, 2005) for its seismic provisions. The seismic provisions of ASCE 7-05 are, in turn, primarily based on the 2003 edition of the *National Earthquake Hazard Reduction Program Recommended Provisions for Seismic Regulations for Buildings and Other Structures* (NEHRP, 2003).

ASCE 7-05 was developed by the ASCE 7 Standards Committee and its Seismic Task Committee. The NEHRP Recommended Provisions were developed by the Building Seismic Safety Council's (BSSC) Provisions Update Committee (PUC) on behalf of the US Department of Homeland

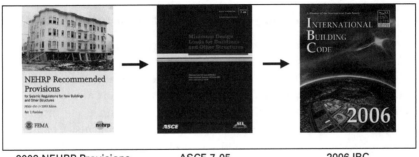

2003 NEHRP Provisions   ASCE 7-05   2006 IBC

1.2 Relationship between documents.

Security's Federal Emergency Management Agency (FEMA). The seismic requirements for nonstructural components were developed by Technical Subcommittee 8 (TS-8) of the PUC. All ASCE and BSSC Committees are purely volunteer activities and are composed of many of the same professionals.

The NEHRP Recommended Provisions were first published in 1985 and have been updated every 3 years since then. The first set of NEHRP Recommended Provisions were based on the *Tentative Provisions for the Development of Seismic Regulations for Buildings*, ATC 3-06 (ATC, 1978) published by the Applied Technology Council (ATC) for the National Bureau of Standards in 1978. This landmark document (one of the first developed by ATC) was prompted by the unexpected poor performance of buildings including nonstructural components (especially hospitals) during the 1971 San Fernando Earthquake. ATC 3-06 has formed the basis for many of the concepts contained in the NEHRP Recommended Provisions and the ASCE 7 standard including those for nonstructural components. Special requirements were included for those components that need to function after design earthquake ground motions that included seismic certification of nonstructural components by shake table tests, experience data, or sophisticated analysis.

The NEHRP Provisions feed directly into the ASCE 7 development process and ASCE 7 in turn serves as a primary referenced standard in the IBC. The seismic design provisions of the 2006 IBC are based on those of ASCE 7-05 and make extensive reference to that standard. In fact, almost all of the seismic design provisions are adopted through reference to ASCE 7-05. The only seismic provisions included in the text of the 2006 IBC are related to ground motion, soil parameters, and determination of seismic design category (SDC), as well as definitions of terms actually used within those provisions and the four exceptions under the scoping provisions. Figure 1.2 illustrates the relationship between the three documents.

1.3   Current building code seismic requirements

The structural requirements of the 2006 IBC, which include seismic requirements, are contained in IBC Chapters 16 to 23. Load combinations and load factors, including those containing seismic loads, are provided in Section 1604 while specific requirements for seismic loads are contained in Section 1613. Chapter 17 contains requirements for testing and inspection including special requirements for nonstructural components. Chapter 18 provides requirements for foundations and Chapters 19 to 23 contain structural element and connection detailing requirements for concrete, aluminum, masonry, steel, and wood. Both the foundation and material requirements have special requirements dealing with seismic loadings.

The seismic requirements found in Section 1613 of the 2006 IBC are rather minimal because of the reliance on referencing ASCE 7-05 seismic provisions. It should be noted that the seismic requirements found in the 2000 and 2003 IBC were much more extensive. The Section 1613 seismic requirements that are provided are as follows:

- General charging language
- Definitions
- Design ground motion parameter definitions and design ground motion maps
- Site soil condition classification definitions and site amplification factors
- Seismic Design Categories based on site ground motions and occupancy
- Reference to ASCE 7-05 for all seismic design criteria requirements
- Two minor alternatives to the ASCE 7-05 seismic design criteria requirements. The first alternative permits structural diaphragms to be assumed to be flexible under certain conditions while the second alternative permits increased height limits for steel ordinary concentrically braced frames and moment frames used in conjunction with seismic isolation systems provided the systems are designed to remain elastic during design earthquake level ground motions.

The ground motion values used for the design of buildings are also used for the design of nonstructural components. Also, the Seismic Design Category that a given building is assigned is one of the key factors that determines the seismic requirements for nonstructural components.

1.3.1   Maximum design earthquake ground motion parameters and ground motion maps

The 2006 IBC defines earthquakes in terms of maximum considered earthquake (MCE) ground motion parameters. The MCE design parameters are defined as those that have a 2% probability of exceedance in 50 years, but with deterministic limits in areas where earthquake sources and return periods are well known. The MCE design parameters are defined for a rock site and are specified as 5% damped spectral ordinates at periods of 0.2 seconds (short period) and at a period of 1.0 second (long period). These spectral values are denoted as S_s and S_1, respectively.

Contour maps developed by the United States Geological Survey (USGS) are provided in Chapter 16 and provide values of S_s and S_1 for all locations in the United States. The values of S_s range from 0.0 to 3.0 g while the values of S_1 range from 0.0 to 1.25 g. Spectral values are also available at a USGS website (http://earthquake.usgs.gov/research/hazmaps/design/index. php), where values are provided based on the latitude and longitude of the site. In areas of high seismicity, where contour values change rapidly, the website is the only accurate way to determine the MCE design parameters.

1.3.2   Site class definitions and site amplification coefficients

It is required by the 2006 IBC to determine the soil profile classification called the Site Class for all new building sites. The Site Class is a function of the soil properties in the top 100 feet of soil at the site. Six Site Classes from A through F are identified by the 2006 IBC, with Site Class A corresponding to very hard rock and Site Class F corresponding to very soft (and possibly liquefiable) soils. Where the soil profile properties at a site are unknown, it is permitted by the 2006 IBC to use Site Class D as the default soil condition unless the building official or geotechnical data determines that Site Class E or F soil is likely to be present at the site (see Section 1613.5.2).

Ground motion site amplification factors are specified in the 2006 IBC in recognition of the significant influence of the soil profile on the earthquake site response. It is also recognized that the magnitude of site amplification for a given soil profile varies with the intensity of the ground motion.

Table 1.1 Site amplification coefficients

Site Class	F_a					F_v				
	S_s					S_1				
	≤ 0.25	0.50	0.75	1.00	≥ 1.25	≤ 0.1	0.2	0.3	0.4	≤ 0.5
A	0.8	0.8	0.8	0.8	0.8	0.8	0.8	0.8	0.8	0.8
B	1.0	1.0	1.0	1.0	1.0	1.0	1.0	1.0	1.0	1.0
C	1.2	1.2	1.1	1.0	1.0	1.7	1.6	1.5	1.4	1.3
D	1.6	1.4	1.2	1.1	1.0	2.4	2.0	1.8	1.6	1.5
E	2.5	1.7	1.2	0.9	0.9	3.5	3.2	2.8	2.4	2.4
F	*	*	*	*	*	*	*	*	*	*

*Site specific determination required.

Factors are provided for both short periods (F_a) and long periods (F_v) in recognition that the amplifications are different in the acceleration-sensitive portion of response as compared to the velocity-sensitive portion of response. The specified factors vary, as shown in Table 1.1.

1.3.3 Design earthquake ground motion parameters

The ground motion parameters that are used for design are called the Design Earthquake Ground Motion Parameters. These parameters are identified as the short period spectral design acceleration S_{DS} and the one second period spectral design acceleration S_{DI}. Both of these design parameters are determined from the MCE parameters and Site Amplification Coefficients as follows:

$$S_{DS} = \tfrac{2}{3} F_a S_s \qquad\qquad\qquad [1.1]$$
$$S_{DI} = \tfrac{2}{3} F_v S_I \qquad\qquad\qquad [1.2]$$

In prior building codes (e.g. the Uniform Building Code), the intent was to design structures for 'life safety' in an earthquake with a 10% probability of being exceeded in 50 years (hereinafter referred to as the 500 year event). However, with the development of the new spectral response acceleration maps, it was recognized that this design basis was not adequate for the infrequent but very large earthquake events that could occur in the eastern United States. Therefore, the decision was made to change the design philosophy so that structures were designed for 'collapse prevention' in a 2500 year event, except for portions of California, where the seismic sources are better known than elsewhere in the country and where a somewhat different approach was used to determine the MCE.

The switch from the 500 year event to the 2500 year event was handled by the new seismic maps. The question became how to accomplish the switch from a 'life-safety' design goal to a 'collapse prevention' design goal. The R modificaton coefficients used in the base shear formula are based on the life-safety goal. The R coefficient is a factor defined in ASCE 7-05 based on the structural system defined for the structure and the seismic detailing associated with that system. It is intended to approximate the adjustment that would need to be made to the elastic response of a given structural system to obtain the probable inelastic response. When considering the question of how to accomplish the 'switch', one approach could have been to adjust the R coefficients to match the collapse prevention goal. However, it was recognized that the R values represented an inherent factor of safety of at least 1.5 (another way of stating it is that structures using the R values could be expected to resist ground shaking, which is 150% of the design level motion before experiencing collapse). Rather than revising the R-

values and perhaps confusing the uninformed users of the code, the code writers opted to reduce mapped spectral response accelerations by the factor of safety of 1.5, which explains the 2/3 factor (2/3 being the reciprocal of 1.5).

Example: Determine the Seismic Design Parameters for the following two building addresses: (1) *334 East Colfax St, Palentine, IL 60067 and (2) 25332 Shadywood Drive, Laguna Niguel, CA 92677*, assuming that the Site Class for the site is 'B'. How do the values change if the Site Class is 'D'?

For Location 1, *334 East Colfax St, Palatine, IL 60067,* the latitude and longitude of the building location are 42.115 and −88.035, respectively. Using the spectral acceleration calculation tool from the USGS website, the mapped MCE spectral accelerations are found to be

$$S_s = 0.165 \text{ and } S_1 = 0.057$$

If Site Class B is assumed, Site Coefficients F_a and F_v are found from Table 1.1 (ASCE 7-05, Tables 11.4-1 and 11.4-2), respectively, as 1.0 and 1.0. Thus, the design spectral accelerations of this building are calculated as

$$S_{DS} = \tfrac{2}{3} F_a S_s = \tfrac{2}{3} \times 1.0 \times 0.165 = 0.11$$

and

$$S_{D1} = \tfrac{2}{3} F_v S_1 = \tfrac{2}{3} \times 1.0 \times 0.057 = 0.038$$

However, if Site Class D is assumed to be at the same building site, then the values of F_a and F_v change to

$$F_a = 1.6 \text{ (for } S_s \leq 0.25)$$
$$F_v = 2.4 \text{ (for } S_1 \leq 0.10)$$

As a result, the design spectral accelerations change to

$$S_{DS} = \tfrac{2}{3} F_a S_s = \tfrac{2}{3} \times 1.6 \times 0.165 = 0.176$$

and

$$S_{D1} = \tfrac{2}{3} F_v S_1 = \tfrac{2}{3} \times 2.4 \times 0.057 = 0.092$$

For Location 2, *25332 Shadywood, Laguna Niguel, CA 92677,* the latitude and longitude of the building location are 33.531 and −117.688, respectively. Using the spectral acceleration calculation tool from the USGS website, the mapped MCE spectral accelerations are found to be

$$S_s = 1.431 \text{ and } S_1 = 0.506$$

If Site Class B is assumed, Site Coefficients F_a and F_v are found from Table 1.1 (ASCE 7-05 Tables 11.4-1 and 11.4-2), respectively, as 1.0 and 1.0. Thus, the design spectral accelerations of this building are calculated as

$$S_{DS} = \tfrac{2}{3} F_a S_S = \tfrac{2}{3} \times 1.0 \times 1.431 = 0.954$$

and

$$S_{D1} = \tfrac{2}{3} F_v S_1 = \tfrac{2}{3} \times 1.0 \times 0.506 = 0.337$$

However, if Site Class D is assumed to be at the same building site, then the values of F_a and F_v change to

$$F_a = 1.0 \, (\text{for } S_s \geq 1.25)$$
$$F_v = 1.5 \, (\text{for } S_1 \geq 0.50)$$

As a result, the design spectral accelerations change to

$$S_{DS} = \tfrac{2}{3} F_a S_s = \tfrac{2}{3} \times 1.0 \times 1.431 = 0.954$$

and

$$S_{D1} = \tfrac{2}{3} F_v S_1 = \tfrac{2}{3} \times 1.5 \times 0.506 = 0.506$$

1.3.4 Occupancy Category and Seismic Design Category

The 2006 IBC requires that building structures be assigned an Occupancy Category and Seismic Design Category. The Occupancy Category of a building structure is based on the Occupancy Category Table 1604.5 in the IBC. In the table, low hazard structures (such as agricultural facilities) are categorized as Occupancy Category I, normal structures are categorized as Occupancy Category II, higher hazard facilities (such as schools) are categorized as Occupancy Category III, and essential facilities (such as hospitals) are categorized as Occupancy Category IV. Figure 1.3 provides a quick summary of the four occupancy categories. The Seismic Design Category (SDC) of a structure is determined from Tables 1.2 and 1.3 (see section 1.4.1 discussion) and is based on the values of S_{DS} and S_{D1} at the structure's site and the Occupancy Category. Figure 1.4 is a flowchart showing how to determine the SDC. The SDC ranges from SDC A to SDC F with SDC A being assigned to the structures in the lowest seismic hazard area and SDC F being assigned to the most essential structures in the highest seismic hazard areas. In general, the higher of the categories in the two tables is the SDC that is assigned to the building structure. There is an exception that permits the SDC to be based only on S_{DS} for certain select short period structures. The Occupancy Category also determines the

Occupancy Category	What is an Occupancy Category I, II, III, IV?
Occupancy Category is a term used to describe the categorization of structures by occupancy or use.	⇒ Occupancy Category I or II is assigned to most buildings
	⇒ Occupancy Category III is for buildings with large numbers of persons such as - Schools with more than 250 students - Assembly uses with more than 300 people - Total occupancy greater than 5,000 people Also non-essential utility facilities Also jails and detention facilities
	⇒ Occupancy Category IV is for hospitals and acute care facilities; fire, police and emergency response stations; and utilities required for essential facilities

1.3 Description of occupancy categories (excerpt taken from Table 1604.5 of the 2006 IBC).

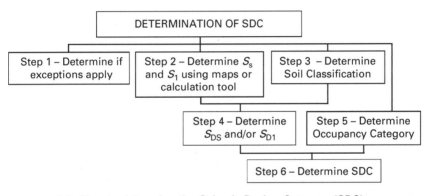

1.4 How to determine the Seismic Design Category (SDC).

building importance factor, I, for the building structure. If the Occupancy Category is I or II, the building importance factor is 1.0. If the Occupancy Category is III, the building importance factor is 1.25, and if the Occupancy Category is IV, the building importance factor is 1.5.

Code-prescribed seismic design is based on the concept of allowing structures that utilize seismic detailing to behave inelastically when subjected to design earthquake ground motions. The concept of increased importance (I) factors for higher performing buildings is linked to the concept that reduced inelastic behavior should result in more reliable structures. Therefore building structures with higher I factors should experience less inelastic deformations (and more reliable performance) than buildings with a similar structural system and lower I factors for the same level of earthquake motions.

Table 1.2 Seismic Design Category based on short period response acceleration (taken from Table 1613.5.6 (1) of the 2006 IBC)

Value of S_{DS}	Occupancy category		
	I or II	III	IV
$S_{DS} < 0.167$ g	A	A	A
0.167 g ≤ $S_{DS} < 0.33$ g	B	B	C
0.33 g ≤ $S_{DS} < 0.50$ g	C	C	D
0.50 g ≤ S_{DS}	D	D	D

Table 1.3 Seismic Design Category based on 1 second response acceleration (taken from Table 1613.5.6(2) of the 2006 IBC)

Value of S_{D1} or S_1	Occupancy category		
	I or II	III	IV
$S_{D1} < 0.067$ g	A	A	A
0.067 g ≤ $S_{D1} < 0.133$ g	B	B	B
0.133 g ≤ $S_{D1} < 0.20$ g	C	C	D
0.20 g ≤ S_{D1}	D	D	D
$S_1 ≥ 0.75$	E	E	F

Example: For the two building sites for which the seismic design parameters were determined previously, determine for the building at each site the Seismic Design Category assuming that the building at each site has normal occupancy (Occupancy Category II), is 10 stories tall, has a period of 1 second, and the soil profile is Site Class D. How would the Seismic Design Category change if the building at each site were an essential facility (Occupancy Category IV)?

For example, Location 1, using Table 1.2, for $S_{DS} = 0.176$ and Occupancy Category II, the Seismic Design Category = B. Using Table 1.3, for $S_{D1} = 0.092$ and Occupancy Category II, the Seismic Design Category = B. Since both tables result in Seismic Design Category B, the building is assigned to SDC B.

For Location 2, using Table 1.2, for $S_{DS} = 0.954$ and Occupancy Category II: Seismic Design Category = D. Using Table 1.3, for $S_{D1} = 0.506$ and Occupancy Category II, the Seismic Design Category = D. Since both tables result in Seismic Design Category D, the building is assigned to SDC D.

If the building was an essential facility (Occupancy Category IV), for Location 1, using Table 1.2, for $S_{DS} = 1.176$ and Occupancy Category IV, the Seismic Design Category = C. Using Table 1.3, for $S_{D1} = 0.092$ and Occupancy Category IV, the Seismic Design Category = C. Since both tables result in Seismic Design Category C, the building is assigned to SDC C.

For Location 2, using Table 1.2, for $S_{DS} = 0.954$ and Occupancy Category IV, the Seismic Design Category = D. Using Table 1.3, for $S_{DI} = 0.506$ and Occupancy Category IV, the Seismic Design Category = D. Since both tables result in Seismic Design Category D, the building is assigned to SDC D.

1.4   2006 IBC/ASCE 7-05 seismic requirements for nonstructural components

Nonstructural components are architectural, mechanical, or electrical components and systems that are permanently attached to structures and that are not considered as part of the primary seismic force resisting structural system. The 2006 IBC references ASCE 7-05 for the seismic design requirements for nonstructural components and their supports and attachments to the primary structure. The seismic requirements for nonstructural components are contained in Chapter 13 of ASCE 7-05. Architectural glass is located in the 'architectural component' section of Chapter 13. Chapter 13 includes the following:

- Scoping language (a.k.a. charging language), importance factor definition, and exemptions
- General design requirements including special seismic certification requirements for designated seismic systems
- Seismic demands on nonstructural components
- Nonstructural component anchorage requirements
- Prescriptive detailing requirements for architectural components
- Prescriptive detailing requirements for mechanical and electrical components.

Nonstructural components are assigned to the same Seismic Design Category as the structure that they occupy, or to which they are attached. In addition to assigning a Seismic Design Category, nonstructural components are also assigned an Importance Factor I_p. The Seismic Design Category of the component and the component importance factor determine the level of prescriptive seismic detailing requirements and special certification requirements. The I_p factor is also considered when determining the design forces on nonstructural components.

1.4.1   Nonstructural component seismic importance factor

The nonstructural component seismic importance factor is taken as 1.0 unless one of the following conditions exists, in which case it is taken as 1.5:

1. The component is required to function after an earthquake for life-safety purposes, including fire protection sprinkler systems.

2. The component contains hazardous materials.
3. The component is in or attached to an Occupancy Category IV building structure (such as a hospital or emergency operations center) and the component is needed for continued operation of the facility, or its failure could impair the continued operation of the facility. Occupancy Category IV buildings have much higher level seismic performance goals for both the structural and nonstructural systems than normal occupancy buildings. For this reason much greater design considerations are required for them.

1.4.2 Designated seismic system

A designated seismic system is defined as a nonstructural system or component that has an importance factor equal to 1.5. This terminology was introduced in ATC 3-06, and its definition has morphed over time. However, the concept of designated seismic systems and the seismic criteria for these systems have not changed. Designated seismic systems are extremely important systems that require special treatment. Examples of designated seismic systems might include fire sprinkler systems, an uninterruptable power supply, and/or emergency power generators. It could also include in hospitals the exterior shell of the building and the heating, ventilating, and air conditioning systems. It is expected that in future editions of ASCE 7, the term 'Designated Seismic Systems' will be replaced simply with 'when $I_p = 1.5$'.

1.4.3 Exemptions

Nonstructural components and their anchorage and bracing can be exempt from the seismic design requirements of Chapter 13 of ASCE 7-05 depending on the Seismic Design Category they have been assigned. Nonstructural components are exempt if one of the following conditions applies:

1. They are assigned to SDC A.
2. They are architectural components assigned to SDC B other than parapets supported by bearing walls or shear walls, provided that $I_p = 1.0$.
3. They are mechanical and/or electrical components assigned to SDC B.
4. They are mechanical and/or electrical components assigned to SDC C, provided that $I_p = 1.0$
5. Mechanical and electrical equipment assigned to SDC D, E or F, provided that $I_p = 1.0$ and one of the two following conditions exist:
   (a) Components weigh 400 pounds or less, are mounted 4 feet or less

above the floor, and there are flexible connections between the components and associated ducting, piping, and bracing.

(b) Components weigh 20 pounds or less or distributed systems weigh 5 lb/foot or less and there are flexible connections between the components and associated ducting, piping, and bracing.

It should be noted that the above exemptions apply, regardless of the type of demand imposed on the component. Therefore architectural glazing is exempt from the seismic requirements of Chapter 13 if it is part of a structure assigned SDC A, or a structure assigned SDC B where the glazing has been assigned an I_p = 1.0.

Example: For the two example building locations evaluated previously in this chapter, determine if building seismic requirements for architectural glazing (which are treated in ASCE 7-05 as architectural components) apply or are exempt.

For building Location 1 with normal occupancy where the SDC is B, by the exemption section of Chapter 13 (as provided above), most architectural components including architectural glazing are exempt from the seismic requirements found in Chapter 13. For this location, if the building has an essential facility occupancy, the assigned SDC would be C, and therefore architectural components, including architectural glazing, are not exempt. Therefore for SDC C, the architectural glazing would be required to satisfy the seismic requirements found in Chapter 13.

For building Location 2, for normal occupancy and essential building, the SDC is D. All architectural components, including architectural glazing, are therefore not exempt and are required to satisfy the seismic requirements found in Chapter 13.

1.4.4 Seismic demands on nonstructural components

In Chapter 13 of ASCE 7-05 two types of nonstructural demands are specified. These are equivalent static lateral forces identified as F_p and relative displacement demands identified as D_p. The F_p forces are at the strength level (i.e. ultimate limit state design) and need to be multiplied by 0.7 when used with allowable stress load combinations and allowable stress increases. The D_p demands are based on the estimated relative displacements of the structure to which they are attached when subjected to design earthquake level demands.

The F_p forces are determined by the following equation:

$$F_p = \frac{0.4a_p S_{DS}}{R_p I_p}\left(1 + 2\frac{z}{h}\right)W_p \qquad [1.3]$$

where

F_p = seismic design force centered at the component's center of gravity and distributed relative to the component's mass distribution
a_p = component amplification factor
S_{DS} = design earthquake spectral response acceleration at short period
R_p = component response modification factor
I_p = component importance factor
Z = height in structure at the point of attachment of the component
H = average roof height of the structure relative to the base elevation
W_p = component operating weight

In addition, the following minimum and maximum values for F_p are specified:

$$F_p \text{ shall not be taken less than } 0.3\,S_{DS}\,I_p\,W_p \qquad [1.4]$$

and

$$F_p \text{ need not be taken greater than } 1.6\,S_{DS}\,I_p\,W_p \qquad [1.5]$$

The values of a_p and R_p are specified in Tables 13.5-1 and 13.6-1 in Chapter 13 of ASCE 7-05 for various nonstructural components (see Tables 1.4 and 1.5). The values of a_p range from 1.0 to 2.5, while the values of R_p range from 1.0 to 12.0. The values found in these tables were established by judgment by the NEHRP TS-8 committee based on the observed behavior and testing of these components subjected to earthquake motions. In many cases they were calibrated to seismic design values for nonstructural components found in previous building codes. The 12.0 value is specified for butt welded steel piping systems because they are expected to exhibit a high degree of inelastic deformation (ductility) before failure. A value of 1.0 is intended for items that would behave in a very brittle fashion before failure. For architectural glazing, the values of a_p and R_p taken from Table 13.5-1 in Chapter 13 are 1.0 and 1.5, respectively (under 'other rigid components, low deformability materials and attachments' entry). The a_p values represent the expected dynamic amplification of flexible components to expected floor motion. If a component is deemed to be rigid (fundamental period of less than 0.06 seconds), no dynamic amplifications need to be considered. In essence, the F_p equation is providing a floor spectra demand on the component, and the component forces can be reduced by an R_p factor in a manner similar to how equivalent forces are determined for building structures. Unlike building structures, however, the forces on nonstructural components are not reduced at longer periods. This is because higher mode effects may result in increased amplification. However, there is an alternate procedure in Chapter 13 that permits the value of F_p to be determined based

Table 1.4 ASCE 7-05 Table 13.5-1 coefficients for architectural components

Architectural component and element	$a_p{}^a$	$R_p{}^b$
Interior nonstructural walls and partitions[b]		
Plain (unreinforced) masonry walls	1.0	1.5
All other walls and partitions	1.0	2.5
Cantilever elements (unbraced or braced to structural frame below its center of mass)		
Parapets and cantilever interior nonstructural walls	2.5	2.5
Chimneys and stacks where laterally braced or supported by the structural frame	2.5	2.5
Cantilever elements (braced to structural frame above its center of mass)		
Parapets	1.0	2.5
Chimneys and stacks	1.0	2.5
Exterior nonstructural walls[b]	1.0^b	2.5
Exterior nonstructural wall elements and connections[b]		
Wall element	1.0	2.5
Body of wall panel connections	1.0	2.5
Fasteners of the connecting system	1.25	1.0
Veneer		
Limited deformability elements and attachments	1.0	2.5
Low deformability elements and attachments	1.0	1.5
Penthouses (except where framed by an extension of the building frame)	2.5	3.5
Ceilings		
All	1.0	2.5
Cabinets		
Storage cabinets and laboratory equipment	1.0	2.5
Access floors		
Special access floors (designed in accordance with Section 13.5.7.2)	1.0	2.5
All other	1.0	1.5
Appendages and ornamentations	2.5	2.5
Signs and billboards	2.5	2.5
Other rigid components		
High deformability elements and attachments	1.0	3.5
Limited deformability elements and attachments	1.0	2.5
Low deformability materials and attachments	1.0	1.5
Other flexible components		
High deformability elements and attachments	2.5	3.5
Limited deformability elements and attachments	2.5	2.5
Low deformability materials and attachments	2.5	1.5

[a]A lower value for a_p shall not be used unless justified by detailed dynamic analysis. The value for a_p shall not be less than 1.00. The value of $a_p = 1$ is for rigid components and rigidly attached components. The value of $a_p = 2.5$ is for flexible components and flexibly attached components. See Section 11.2 for definitions of rigid and flexible.
[b]Where flexible diaphragms provide lateral support for concrete or masonry walls and partitions, the design forces for anchorage to the diaphragm shall be as specified in Section 12.11.2.

Table 1.5 ASCE 7-05 Table 13.6-1 seismic coefficients for mechanical and electrical components

Mechanical and electrical components	$a_p{}^a$	$R_p{}^b$
Air-side HVAC, fans, air handlers, air conditioning units, cabinet heaters, air distribution boxes, and other mechanical components constructed of sheet metal framing	2.5	6.0
Wet-side HVAC, boilers, furnaces, atmospheric tanks and bins, chillers, water heaters, heat exchangers, evaporators, air separators, manufacturing or process equipment, and other mechanical components constructed of high-deformability materials	1.0	2.5
Engines, turbines, pumps, compressors, and pressure vessels not supported on skirts and not within the scope of Chapter 15	1.0	2.5
Skirt-supported pressure vessels not within the scope of Chapter 15	2.5	2.5
Elevator and escalator components	1.0	2.5
Generators, batteries, inverters, motors, transformers, and other electrical components constructed of high-deformability materials	1.0	2.5
Motor control centers, panel boards, switch gear, instrumentation cabinets, and other components constructed of sheet metal framing	2.5	6.0
Communication equipment, computers, instrumentation, and controls	1.0	2.5
Roof-mounted chimneys, stacks, cooling and electrical towers laterally braced below their center of mass	2.5	3.0
Roof-mounted chimneys, stacks, cooling and electrical towers laterally braced above their center of mass	1.0	2.5
Lighting fixtures	1.0	1.5
Other mechanical or electrical components	1.0	1.5
Vibration isolated components and systems^b		
Components and systems isolated using neoprene elements and neoprene isolated floors with built-in or separate elastomeric snubbing devices or resilient perimeter stops	2.5	2.5
Spring isolated components and systems and vibration isolated floors closely restrained using built-in or separate elastomeric snubbing devices or resilient perimeter stops	2.5	2.0
Internally isolated components and systems	2.5	2.0
Suspended vibration isolated equipment including in-line duct devices and suspended internally isolated components	2.5	2.5
Distribution systems		
Piping in accordance with ASME B31, including in-line components with joints made by welding or brazing	2.5	12.0
Piping in accordance with ASME B31, including in-line components, constructed of high or limited deformability materials, with joints made by threading, bonding, compression couplings, or grooved couplings	2.5	6.0
Piping and tubing not in accordance with ASME B31, including in-line components, constructed of high-deformability materials, with joints made by welding or brazing	2.5	9.0
Piping and tubing not in accordance with ASME B31, including in-line components, constructed of high- or limited-deformability materials, with joints made by threading, bonding, compression couplings, or grooved couplings	2.5	4.5

Table 1.5 (cont.)

Mechanical and electrical components	$a_p{}^a$	$R_p{}^b$
Piping and tubing constructed of low-deformability materials, such as cast iron, glass, and nonductile plastics	2.5	3.0
Ductwork, including in-line components, constructed of high-deformability materials, with joints made by welding or brazing	2.5	9.0
Ductwork, including in-line components, constructed of high or limited-deformability materials with joints made by means other than welding or brazing	2.5	6.0
Ductwork, including in-line components, constructed of low-deformability materials, such as cast iron, glass, and nonductile plastics	2.5	3.0
Electrical conduit, bus ducts, rigidly mounted cable trays, and plumbing	1.0	2.5
Manufacturing or process conveyors (nonpersonnel)	2.5	3.0
Suspended cable trays	2.5	6.0

[a]A lower value for a_p is permitted where justified by detailed dynamic analyses. The value for a_p shall not be less than 1.0. The value of a_p equal to 1.0 is for rigid components and rigidly attached components. The value of a_p equal to 2.5 is for flexible components and flexibly attached components.
[b]Components mounted on vibration isolators shall have a bumper restraint or snubber in each horizontal direction. The design force shall be taken as $2F_P$ if the nominal clearance (air gap) between the equipment support frame and restraint is greater than 0.25 in. If the nominal clearance specified on the construction documents is not greater than 0.25 in, the design force is permitted to be taken as F_P.

on a dynamic analysis where the building and nonstructural component are analyzed together in a single model.

The F_p force is primarily used for design anchorage and bracing of the component. For architecture glazing, the F_p force is typically applied out-of-plane for purposes of checking the glass panel frame connection. It is presumed for glazing that the in-plane glazing capacity has already been checked by testing or by providing adequate clearances around the edges of the glazing, as discussed below.

The relative displacement demand D_p is determined from the analysis of the structure in which the components are being attached. The analysis displacements need to include the deflection amplification factor C_d. As a default, if the relative displacements are unknown, the relative displacement demands may be taken as the maximum allowable drift displacements allowed for the structure by ASCE 7-05 as follows:

$$D_p = \delta_{xA} - \delta_{yA} \qquad [1.6]$$

except that D_p need not be taken greater than

$$D_p = (h_x - h_y)\Delta_{aA}/h_{sx} \qquad [1.7]$$

Table 1.6 ASCE 7-05 Table 12.12-1 allowable story drift, $\Delta_a{}^{a,b}$

	Occupancy category		
Structure	I or II	III	IV
Structures, other than masonry shear wall structures, 4 stories or less with interior walls, partitions, ceilings, and exterior wall systems that have been designed to accommodate the story drifts	$0.025h_{sx}{}^c$	$0.020h_{sx}$	$0.015h_{sx}$
Masonry cantilever shear wall structures^d	$0.010h_{sx}$	$0.010h_{sx}$	$0.010h_{sx}$
Other masonry shear wall structures	$0.007h_{sx}$	$0.007h_{sx}$	$0.007h_{sx}$
All other structures	$0.020h_{sx}$	$0.015h_{sx}$	$0.010h_{sx}$

^{a,b,c,d} See ASCE 7-05 Table 12.12-1 for footnotes.

where

D_p = relative seismic displacement that the component must be designed to accommodate

δ_{xA} = deflection at building level x determined by elastic analysis, which includes the C_d multiplier but excludes the I factor for the building

δ_{yA} = deflection at building level y determined by elastic analysis, which includes the C_d multiplier but excludes the I factor for the building

h_x = height of level x where the upper connection point is attached

h_y = height of level y where the lower connection point is attached

h_{sx} = story height of the structure used to determine allowable story drift in Chapter 12 of ASCE 7

Δ_{aA} = allowable story drift of the structure as defined in Chapter 12 of ASCE 7 (see Table 1. 6)

The relative displacement demand is used to determine the effects on displacement sensitive components caused by relative anchor movements. For such components inelastic deformations are acceptable, but failure of the component, which can cause life-safety hazard or loss of function (if required for essential operations), is not.

Example: For the two example building locations evaluated earlier, determine the relative displacement demand, D_p, that would need to be accommodated by architectural glazing that is attached vertically at the floor levels assuming that the buildings are 10 stories high, individual story heights are 4100 mm (13.45 feet), the buildings are not constructed of masonry, and structural analysis results for the buildings are not available.

The default option for determining the relative displacement demand for architectural glazing is based on Equation [1.7] and the allowable story drift is specified in ASCE 7-05 Table 12.2-1 (presented as Table 1.6). It should be noted that allowable story drift is a function of only the structural system type, number of stories, and structural occupancy. It is not a function of location or design ground motion level. However, the seismic design exemptions discussed earlier are a function of the design ground motion level at a given location.

For both example building locations where the building is normal occupancy (Occupancy Category II), not constructed of masonry, and over four stories in height, the allowable story drift, Δ_{aA}, is 0.020 h_{sx}. If the same building were an essential facility (Occupancy Category IV), the allowable story drift, Δ_{aA}, is 0.010 h_{sx}. For glazing in an essential facility, the default glazing relative displacement demand is scaled by an I factor of 1.5, so the default drift demand margin is 0.015 h_{sx}. This seems like a paradox since the more critical facility is designed for less displacement demand. In reality it is not, since the building structural frame is required to limit drift deflections to 0.010 h_{sx} and thus there is 50% more glazing gap margin for the essential facilities than for normal buildings. Since the intention of the more restrictive drift limit for higher occupancy buildings was to improve nonstructural performance using normal occupancy installations, it is recommended that a normal occupancy glazing design which has a drift capacity of 0.020 h_{sx} be used in all circumstances.

Therefore, for building Locations 1 and 2, for normal occupancy (Occupancy Category II) where the points of attachments are at the floor levels, the displacement demand, D_p, is determined by substituting into Equation [1.7] as follows:

$$D_p = (h_x - h_y)\Delta_{aA}/h_{sx} = (4100\,\text{mm})\,0.020\,(4100\,\text{mm})/(4100\,\text{mm})$$
$$= 0.020\,(4100\,\text{mm})$$
$$= 82\,\text{mm}\,(3.2\,\text{in})$$

For building Locations 1 and 2, for essential occupancy (Occupancy Category IV) where the points of attachments are at the floor levels, the displacement demand, D_p, is determined by substituting into Equation [1.7] as follows:

$$D_p = (h_x - h_y)\Delta_{aA}/h_{sx} = (4100\,\text{mm})\,0.010\,(4100\,\text{mm})/(4100\,\text{mm})$$
$$= 0.010\,(4100\,\text{mm})$$
$$= 41\,\text{mm}\,(1.6\,\text{in})$$

It is recommended using a design that can accommodate 82 mm (3.2 in).

1.4.5 Special requirements for designated seismic systems

Designated seismic systems are expected to have a higher likelihood of functioning following design earthquake level motions. To improve the probability of achieving this expectation, Chapter 13 of ASCE 7-05 has more stringent requirements for these systems. These requirements include designing anchorage, bracing, and the component itself for higher-level seismic forces and explicitly evaluating seismic anchor movement effects. In addition, special certification is required, indicating that the components will perform as intended if the components are assigned Seismic Design Category C to F. Certification is required for:

- Mechanical and electrical equipment are required to demonstrate and certify compliance by either shake table testing or experience data.
- Components containing hazardous material are required to demonstrate and certify compliance by shake table testing, experience data, or analysis.

In addition, Chapter 13 requires consequential damage considerations such that failure of nonessential nonstructural components will not cause the failure of essential nonstructural components.

1.5   Seismic requirements for architectural glass

In the 2000 NEHRP Recommended Provisions, seismic requirements for architectural glazing were first introduced. These requirements were based on recommendations by Richard Behr based on testing performed for the National Science Foundation and the glazing industry at Penn State University, whose results were provided in the Commentary to the 2000 NEHRP Recommended Provisions (NEHRP, 2000). It should also be stated that the interdisciplinary group of experts on NEHRP TS-8, reviewed, deliberated, and approved these recommendations. The recommendations were then reviewed, deliberated, and approved by the NEHRP Provisions Update Committee and the membership of the Building Seismic Safety Council. The seismic requirements were subsequently adopted into ASCE 7-02 and, by reference, were adopted into the 2003 IBC. The same requirements are found in ASCE 7-05 and, for reference, the 2006 IBC.

Glass in glazed curtain walls, storefronts, and partitions can be sensitive to in-plane relative displacements (i.e. horizontal racking displacements). Therefore, the seismic requirements are a function of the relative displacement demand D_p and the glass relative displacement capacity $\Delta_{fallout}$. The seismic requirements found in Section 13.5.9 of ASCE 7-05 for architectural glass are as follows:

$$\Delta_{fallout} \geq 1.25ID_p \qquad [1.8]$$

or 0.5 inches (13 mm) whichever is greater, where Δ_{fallout} is the relative displacement capacity of the glass system, defined in Section 13.5.9 of ASCE 7-05 as the drift that causes glass to fall out of its framing as determined by the American Architectural Manufacturers Association (AAMA) Standard 501.6, *Recommended Dynamic Test Method for Determining the Seismic Drift Causing Glass Fallout from a Wall System* (AAMA, 2001), or as determined by engineering analysis. I is the importance factor for the building (see Section 1.3.4 of this chapter) and D_p is the relative displacement demand (see Section 1.4.4 of this chapter).

It should be noted that the seismic requirement for glass only applies to buildings assigned to Seismic Design Category C, D, E, or F. If any of the following three conditions exist, the glass system is exempt from the seismic requirements:

1. There are adequate clearance gaps around the glass such that physical contact will not occur between the outer edge of the glass and the framing system at the design drift D_p. The concept for this requirement is that if adequate clearance exists, which is the physical clearance around the edge of the glass plus a margin, then the glass is expected to remain uncracked. Adequate clearance is defined as when D_{clear} is greater than 1.25 times D_p, where D_{clear} is defined for rectangular glass panels as

$$D_{\text{clear}} = 2c_1\left(1 + \frac{h_p c_2}{b_p c_1}\right)$$
[1.9]

where

D_{clear} = the relative horizontal (drift) displacement, measured over the height of the glass panel under consideration, which causes initial glass-to-frame contact simultaneously at opposing diagonal corners of an architectural glass panel within a dry-glazed curtain wall frame

h_p = the height of the rectangular glass panel

b_p = the width of the rectangular glass panel

c_1 = the clearance (gap) between the vertical glass edges and the frame

c_2 = the clearance (gap) between the horizontal glass edges and the frame

From a conceptual viewpoint, for this exception, the relative *demand* on the glazing is 1.25 D_p while the displacement *capacity* of the glazing is D_{clear}. The above terms are illustrated in Fig. 1.5.

2. Fully tempered monolithic glass in Occupancy Category I, II, and III buildings located no more than 10 feet (3 m) above a walking surface.

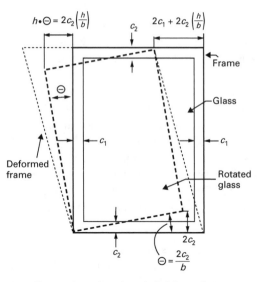

1.5 Illustration showing definition of clearance terms (by A. Memari, 2008).

3. Annealed or heat-strengthened laminated glass with an interlayer no less than 0.030 inches (0.76 mm) that is captured mechanically in a wall system glazing pocket and whose perimeter is secured to the frame by a wet glazed system gunable curing elastomeric sealant perimeter bead of 0.5 inches (13 mm) minimum glass contact width, or other approved anchorage system.

Example: For the two example building locations evaluated previously in this chapter, provide architectural glazing design options that satisfy the seismic requirements of Chapter 13 of ASCE 7-05, assuming that the building has normal occupancy (Occupancy Category II).

Note that the following example was taken from Behr (2006).

*   Glass panel dimensions are 1524 mm (five feet) wide by 1829 mm (six feet) high.
*   Dry-glazed (fixed rubber gasket) aluminum curtain wall system.
*   Clearance between each glass edge and the aluminum wall frame glazing pocket is 10 mm (3/8 inches).
*   Glass panel has side blocks at mid height along both vertical glass edges and is supported by rubber setting blocks at the quarter points along the bottom horizontal glass edge.

Glazing system data (see Fig. 1.6)

Given

Unless there is a special, movement-accommodating connection between the main structural system and the curtain wall framing system, then all the story drifts calculated for the main structural system are assumed to be transferred to the curtain wall framing system because of continuous vertical mullions over more than one story. In other words, the curtain wall is assumed to be completely coupled to the main building structural system.

The ASCE 7-05 seismic design equation to be satisfied for the selected glass panel is (from ASCE 7-05 Section 13.5.9.1)

$$\Delta_{\text{fallout}} \geq 1.25 \times I \times D_p \qquad [1.10]$$

Here, the occupancy importance factor $I = 1.0$ and D_p is the relative seismic displacement that the glass panel must be designed to accommodate, taken over the height of the glass panel:

$$D_p = \frac{\text{glass panel height}}{\text{story height}} \times \text{story drift} = \frac{1829\,\text{mm}}{4100\,\text{mm}} \times 82\,\text{mm}$$
$$= 36.6\,\text{mm}\,(1.44\,\text{in})$$

With a glass and aluminum curtain wall system, it is appropriate to assume that story drift is evenly distributed over the story height. (In other situations, such as window panels surrounded by stiff, precast concrete panels above and below the window panel, it would be appropriate to assume that 100% of the design story drift must be accommodated over the height of the glass panel, which would impose twice the drift demand on the glass as compared to the value used in this design example.) Therefore,

$$\Delta_{\text{fallout}} \geq 1.25 \times I \times D_p = 1.25 \times 1.0 \times 36.6\,\text{mm} = 45.8\,\text{mm}(1.80\,\text{in})$$

Design options to satisfy the Δ_{fallout} requirement

1. Test a mock-up of the curtain wall constructed with an economical glass type (one that also satisfies the wind load structural requirements for the building being designed) in accordance with AAMA 501.6 at an AAMA-accredited testing laboratory. If Δ_{fallout} obtained by the AAMA 501.6 test method for the trial design is greater than 45.8 mm, then the trial design is acceptable.
2. In accordance with Exception 1 in Section 13.5.9.1 of ASCE 7-05, provide sufficient glass-to-aluminum frame (mullions) edge clearance to avoid contact at the design drift:

$$D_{\text{clear}} \geq 1.25 D_p \qquad \text{(ASCE 7-05 Equation 13.5 − 2)}$$

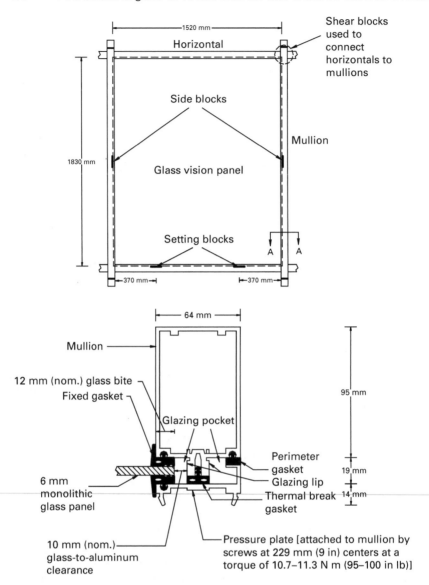

1.6 Glazing details for the seismic design example.

where

$$D_{\text{clear}} = 2c_1\left(1 + \frac{h_p c_2}{b_p c_1}\right)$$

(definitions of equation parameters were given earlier in this chapter). Here

$$D_{clear} = 2(10)\left[1 + \frac{1829(10)}{1524(10)}\right] = 44\,mm\,(1.73\,in)$$

D_{clear} = 44 mm is the amount of horizontal racking displacement (applied over the height of the glass panel) required to cause initial glass-to-frame contact simultaneously at opposing corners of a main diagonal of the rectangular glass panel.

Here, D_{clear} = 44 mm is less than the required $1.25 \times D_p$ = 1.25 × 36.6 = 45.8 mm (1.80 in), so diagonal glass-to-frame contact is *not* avoided in the present trial design. (The glass-to-frame clearances could be increased to, say, 13 mm (1/2 in) to satisfy ASCE 7-05 Equation 13.5-2 and provide an acceptable design in this example, i.e. 57.2 mm (2.25 in).)

3. In accordance with Exception 2 in Section 13.5.9.1 of ASCE 7-05, because this building is categorized as Occupancy Category II, fully tempered monolithic glass may be used for glazing panels no more than 10 feet (3 m) above a walking surface. (This exception to Section 13.5.9.1 would have only limited applicability to most multistory building wall system designs.)
4. In accordance with Exception 3 in Section 13.5.9.1 of ASCE 7-05, annealed or heat-strengthened laminated glass in single thickness with an interlayer no less than 0.030 inch (0.76 mm) that is captured mechanically in a wall system glazing pocket, and whose perimeter is secured to the frame by a wet-glazed gunable curing elastomeric sealant perimeter bead of 1/2 inch (13 mm) minimum contact width, or other approved anchorage system, would provide an acceptable glazing system design for glass fallout resistance.

Design example end note

In a situation involving a higher drift demand (D_p) on the glazing system versus the value used in this example, selecting a preliminary design that is more inherently glass fallout resistant (e.g. annealed laminated glass units, heat-strengthened laminated glass units, a wider mullion design with larger glass-to-frame clearances, etc.) and testing a representative wall system mockup in accordance with AAMA 501.6 at an AAMA-certified testing laboratory, would be an appropriate course of action. If $\Delta_{fallout}$ obtained by the AAMA 501.6 test method for the trial design exceeds the required $\Delta_{fallout}$ from Equation [1.10], then the trial design is acceptable.

1.6   Future trends

While the seismic requirements for glass have been in building codes since the adoption of the 2003 IBC, past experience indicates that the real implementation of seismic requirements in US building codes normally does not occur until California adopts and enforces such building code requirements. California has typically taken the lead on code enforcement of seismic requirements and other states typically follow the lead of California's enforcement interpretations as appropriate. However, California did not adopt the 2003 IBC and did not begin enforcing the 2006 IBC until January 2008. The reason that California did not adopt the IBC sooner was purely political and is associated with a rival building code developed by the National Fire Protection Association. The former governor of California supported the rival code. However, that governor fell out of favor with the citizens of California, and was recalled and replaced by the current governor, Arnold Schwarzenegger. The current governor replaced the commission membership that decides building codes in California and the new commission decided to adopt the IBC starting with the 2006 edition. Therefore, while the code seismic requirements for glass have existed for some time, how glazing designs will change to accommodate the new requirements will not be seen for a few years. It is anticipated that the glazing industry will react by modifying both designs and testing requirements, and refined code requirements could also result.

Another future trend is performance-based earthquake engineering. The Applied Technology Council (ATC) is currently developing the Next-Generation Performance Based Earthquake Engineering Guidelines. This effort is called the ATC-58 Project and is sponsored by the US Department of Homeland Security's Federal Emergency Management Agency (FEMA). The project is utilizing an approach where both demands and capacities are defined by probabilistic formulations. Performance goals will consider the full range of possible input motions and the resulting damage and losses that might occur. The goals will focus on the consequences, including casualties, economic loss, and downtime associated with operability and repairs. Capacities will be established on a probabilistic basis and will consider a range of possible damage states. This means that for glazing, capacities will be established as fragilities for various types of glazing systems and the variations of installation conditions.

It is expected that these fragilities will utilize interstory drifts as the demand parameters and that the damage states will not only include glass fallout, but glass breakage and seal damage. It is not expected that these next generation, performance-based earthquake engineering procedures will be adopted into model building codes in the United States for at least 10 to 20 years. Other than providing drift requirements for glazing systems, it does not appear at this time that other countries are pursuing performance-based seismic requirements such as these for glazing. Therefore, developments in the United States could receive considerable international attention.

1.7   Sources of further information and advice

The reader is directed to the commentary of the 2003 NEHRP Recommended Provisions (FEMA 450-2) for further background on the architectural glass seismic requirements that are currently found in the 2003 NEHRP Recommended Provisions and in ASCE 7-05. Background and details on the seismic requirements for glazing and AAMA testing can also be found in Behr (2006). The reader is also directed to the commentary of the 2009 NEHRP Recommended Provisions, which is intended to be a commentary on the seismic requirements of ASCE 7-05. Both the provisions and commentary documents are available at no cost from FEMA (at http://www.fema.gov/library/viewRecord.do?id = 2744) or from BSSC (at www.bssconline.org). Background materials and reports on the ATC-58 project can be found at the ATC website (www.atcouncil.org).

1.8   References

AAMA (2001) *Recommended Dynamic Test Method for Determining the Seismic Drift Causing Glass Fallout from a Wall System*, standard 501.6, American Architectural Manufacturers Association, Schaumburg, IL.

ASCE (2005) *Minimum Design Loads for Buildings and Other Structures*, ASCE 7-05, American Society of Civil Engineers, Reston, VA.

ATC (1978) *Tentative Provisions for the Development of Seismic Regulations for Buildings*, ATC 3-06, Applied Technology Council, Washington, D.C.

Behr, R. A. (2006) Design of architectural glazing to resist earthquakes, *Journal of Architectural Engineering – Special Edition*, 12 (3), 122-128, American Society of Civil Engineers, Reston, VA.

IBC (2006) *International Building Code*, International Code Council, Falls Church, VA.

Memari, A. M., O'Brien, W.C., Kremer, P.A. and Behr, R.A. (2008) *Architectural Glass Seismic Behavior Fragility Curve Development, Final Report for the ATC-58 Project*, Applied Technology Council, Redwood City, CA.

NEHRP (2003) *National Earthquake Hazard Reduction Program Recommended Provisions for Seismic memaRegulations for Buildings and Other Structures*, Federal Emergency Management Agency, Washington, DC.

2

Glazing and curtain wall systems to resist earthquakes

A . M . M E M A R I , The Pennsylvania State University, USA
and T. A. S C H W A R T Z , Simpson Gumpertz & Heger, USA

Abstract: This chapter reviews lessons learned from the seismic response of glass in windows, storefronts, and curtain walls in actual earthquakes, as well as in laboratory experimental studies involving in-plane cyclic racking tests. New developments in analytical methods to predict the response of architectural glass in earthquakes (e.g. drift that causes glass cracking) are discussed. The authors also discuss methods to mitigate architectural glass damage during earthquakes, including the use of sufficient glass-to-frame clearance, fully tempered and/or laminated glass, structural silicone glazing (SSG), adhered and anchored safety films, and rounded corner glass.

Key words: glazing systems seismic behavior, laboratory cyclic racking tests, analytical glass damage prediction, seismic damage mitigation.

2.1   Introduction

Recent earthquakes have revealed the vulnerability of glazing systems to seismic damage according to reconnaissance reports (EERI, 1990, 1995a, 1995b, 2001). These documents confirm that earthquake damage has occurred in wall systems containing architectural glass components on buildings that have experienced little or no damage to the primary structural system (FEMA, 1994). Failure of a component of curtain wall (CW) and storefront systems can pose life-safety hazards due to falling glass or other cladding materials.

In general, contemporary building envelope wall systems are regarded as 'nonstructural components' because they are normally not designed to contribute to the load-carrying capability of the building structural systems. The performance of buildings in past earthquakes shows that building envelope components sustain damage. The damage to these nonstructural components is usually the result of an incompatibility between the deformation characteristics of the structural framing and the movement capability of the cladding, e.g. insufficient perimeter joint widths and lack of slip-accommodating connections.

In response to concerns about nonstructural components damage during earthquakes, recent seismic provisions (e.g. ASCE 7-05) adopted by model building codes, e.g. International Building Code 2006 (ICC, 2006), require nonstructural architectural components that could pose a life-safety hazard to be designed to accommodate the seismic relative displacement requirements determined as design building story drifts. ASCE 7-05 seismic provisions reference the American Architectural Manufacturer's Association (AAMA) test procedure (AAMA, 2001) for determining the glass fallout resistance from CW and storefront wall system mock-ups, and provide design guidance for the acceptable seismic performance of such wall systems. Recent developments in seismic codes require that curtain walls be able to absorb greater degrees of drift than in the past (Bell and Zarghamee, 2004).

A series of laboratory studies and some post-earthquake reconnaissance surveys during the last twenty years have generated a substantial database of information on the expected seismic performance of various combinations of architectural glass and wall framing (Behr, 1998; Behr and Belarbi, 1996; Behr et al., 1995a, 1995b; Deschenes et al., 1991; EERI, 1995a, 1995b, 2001; Evans et al., 1988; King and Thurston, 1992; Lim and King, 1991; Lingnell, 1994; Memari et al., 2003; Pantelides and Behr, 1994; Thurston and King, 1992; Wang et al., 1992; Wright, 1989). Additional studies have also been directed toward the development of seismic isolation methods for new wall system installations and techniques to predict and mitigate glass damage and glass fallout for existing wall systems (Brueggeman et al., 2000; Memari et al., 2004, 2006c; Zarghamee et al., 1995).

This chapter reviews lessons learned from the seismic response of glass in windows, storefronts, and CWs in actual earthquakes and also in laboratory experimental studies. The efforts made to mitigate architectural glass damage are also reviewed. Damage to glass in past earthquakes and forensic studies offer lessons that can be used to improve the behavior of architectural glass in such events. Laboratory studies undertaken to develop a better understanding of glass behavior under simulated seismic conditions enhances the lessons learned. A brief review of earlier experimental studies and a more detailed review of a recent study on conventional glazing systems and the behavior of insulating glass (IG) units subjected to in-plane cyclic racking are also presented.

The last part of the chapter focuses on recent developments in research and products to improve glazing system behavior in earthquakes and identifies relevant R&D areas for follow-up efforts. One study relates to the response of CW mock-ups glazed with annealed monolithic architectural glass panels fitted with an adhered and mechanically anchored polyethylene terephthalate (PET) film under simulated earthquake conditions. Another study that is reviewed is a recently developed seismic damage mitigation concept, which involves using architectural glass panels with modified corner geometries and edge finish conditions to improve their resistance to earthquake damage. The last study is an experimental research program on simulated seismic performance of structural silicone glazed CW systems, including performance of a full-scale, two-sided structural silicone glazing mock-up made up of three side-by-side glass lites or panels tested under cyclic racking displacements to determine serviceability and ultimate behavior responses. The chapter concludes with remarks on current architectural and structural engineering trends related to glazing systems that require additional research for earthquake safety.

2.2   Types of glazing and curtain wall systems

This section presents the attributes of several commonly used glazing systems employing architectural glass, including punched window systems, strip window systems, storefronts, and CW systems. All these systems may be parts of the building envelope, which has several functions including protection of the interior against environmental loads (e.g. air, thermal, moisture) and structural loads (e.g. gravity, wind, seismic). In general, glazing systems are 'non-load bearing' components, meaning that they do not carry the floor gravity loads, except their own self-weight, nor do they usually participate in the overall lateral-load resistance of the structural frame. They do, however, transfer out-of-plane wind load and seismic loads to the main structural system. Glass thickness is based on building code specified wind loads that cause bending in the glass (ASTM E 1300-04, 2004a; GANA, 2004).

Glass is 'captured' by the framing through mechanical means, i.e. a combination of removable and fixed stops on the inside or outside of the frame to hold the glass in place. For conventional 'captured' glazing, in which the glass is held by an external mechanical stop, the face clearance between the stop and the glass is filled with a soft material that cushions the glass in its fixture and forms the air and water seal between the glass and the frame. The filler can be a rubber gasket (dry gasket glazing), as shown in Fig. 2.1, or a liquid-applied sealant (wet glazing). An alternative to captured glazing is structural silicone glazing (SSG), which bonds the glass to the frame through adhered silicone sealant, as shown in Fig. 2.2. Figures 2.3 and 2.4 show, respectively, photographs of buildings with typical dry-glazed and structural silicone-glazed curtain walls.

Punched windows are generally surrounded by other wall cladding materials on all four sides, as shown in Fig. 2.5. This configuration requires

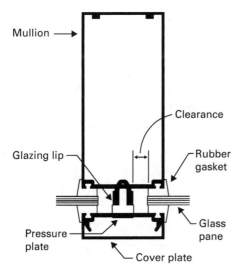

2.1 Typical section of a dry-glazed curtain wall frame system.

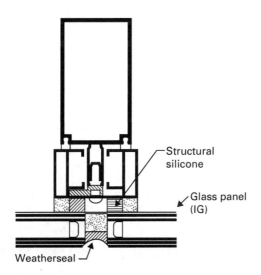

2.2 Typical detail for a structural silicone glazing (SSG) curtain wall configuration.

deformation compatibility under load between the window and the surrounding wall elements.

Strip windows are more common in office buildings and normally span vertically between horizontal spandrel beams and panels, as shown in Fig. 2.6. Typically, the glazing frame will experience essentially all of the building's full interstory drift as the upper spandrel or edge beams move with respect to the lower one. However, if the glazing frame connection to

2.3 Typical dry-glazed curtain wall system.

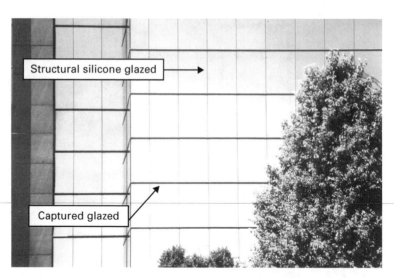

Structural silicone glazed

Captured glazed

2.4 Typical 'two-sided' structural silicone glazed curtain wall system (i.e. verticals are structurally glazed; horizontals are conventional 'captured' glazing).

the (top) spandrel is isolated properly from the spandrel through slip joints, the glazing frame will not experience the drift-induced deformation.

Curtain walls, as the name implies, are 'hung' on the exterior of the building as a curtain (Figs 2.3 and 2.4), transferring lateral (wind and seismic inertia) loads imposed on them to the supporting building frame and supporting only their own weight. In glass/metal CWs, the glazing frame is

2.5 Typical punched window glazing system.

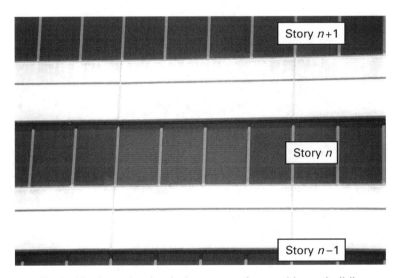

2.6 Typical horizontal strip glazing system in a multistory building.

usually attached to the structural frame through clip angles and restrained from out-of-plane movement at each story. In these systems, the mullion, which may be continuous over two or more stories, as shown in Fig. 2.7, must deform to accommodate building interstory drifts. Figure 2.7 shows examples of mullion attachments to a structural steel frame and concrete floor system. If the curvature resulting from such lateral deformation consumes the glass-to-metal clearance (defined in Fig. 2.1) built into the glazing system, as shown also in Fig. 2.8, the glass will come into contact with

2.7 Typical attachment through clip angles of mullions to steel frame and concrete floor over two adjacent stories.

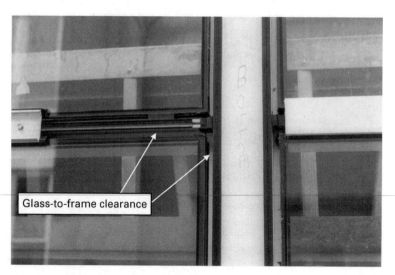

2.8 Typical glass-to-frame clearance.

the frame, and, therefore, likely be damaged and break. Bending curvature of the mullions across the spandrel line may also break spandrel glass.

Some CWs are stick built, meaning the glazing frames are assembled in place from individual members that are anchored to the structural system at each floor level. An alternative system is called unitized, consisting of generally pre-glazed, pre-assembled subframes that are erected into a mullion framing system.

The glass lites infilling the curtain wall frames may be annealed (AN), heat-strengthened (HS), or fully tempered (FT), and can be monolithic, or combined to form laminated and insulating glass (IG) units. Heat treatment of glass (i.e. HS or FT glass) (ASTM C 1048-04, 2004b) creates residual surface compressive stress in the range of 24.13 MPa (3500 psi) to 51.71 MPa (7500 psi) for HS, or over 68.95 MPa (10 000 psi) for FT (ASTM E 1300-04, 2004a) in the glass, which increases resistance to fracture under bending stress. The surface residual stress makes HS and FT glass about two times and four times stronger, respectively, in out-of-plane flexure than the same size and thickness of AN glass (GANA, 2004).

Storefront glazing systems are similar to strip window systems in that they generally span between floors, i.e. supported at or near the first floor, and span to the second floor as shown in Fig. 2.9. Storefront glass lites are usually larger than lites in strip windows.

Glass breakage during seismic events generally occurs from glass-to-metal contact and resulting damage to the glass edge weakens the glass. Such contact occurs because the clearance between the glass and the framing is consumed by the lateral deformation of the frames. In SSG configurations, the glass generally does not contact a framing member because it is isolated from the framing by the silicone seals (Fig. 2.2), but the excessive deformation of the supporting frames can result in loss of structural sealant adhesion and loss of attachment of whole lites of glass. Reduction of adhesive capacity of the sealant is also possible under out-of-plane deflection of the SSG panel under cyclic loads (La Tona *et al.*, 1988). Figure 2.10 shows a typical gap between the adjacent edges of glass panels

2.9 Example of storefront glazing.

Gap between edges
of glass panels

2.10 Gap between the edges of glass panels in two-sided SSG
configuration (i.e. the vertical joint is SSG).

(in the case shown, an IG unit) in an SSG configuration. The gap is filled
with a silicone weatherseal over a backer rod, as shown in Fig. 2.2.

2.3   Performance of glazing and curtain wall systems in past earthquakes

US earthquakes during the past 10–15 years, e.g. 1989 Loma Prieta (EERI,
1990), 1994 Northridge (EERI, 1995a), and 2001 Nisqually (EERI, 2001),
have revealed the vulnerability of modern building envelopes. Outside the
US, significant damage to building envelope components has occurred in
Japan earthquakes, e.g. 1995 (Kobe) Hyogoken-Nambu (EERI, 1995b;
FGMAJ, 1995). In general, damage to glazing systems in earthquakes has
included glass fracture, excessive and permanent deformation of the glazing
frame, and loss of attachment in adhered glass systems. Such damage poses
life-safety hazards as shards of glass have fallen on to sidewalks and streets.
Figures 2.11 to 2.16 show some photographs of glass damage in past
earthquakes.

The damage in recent earthquakes shows that even under minor and
moderate events, glass breakage can occur. The earthquake of July 20, 2007
with a magnitude of 4.2 in the San Francisco bay area caused extensive glass
damage to some buildings (SFGATE, 2007). Figure 2.16 shows the
aluminum mullion attached to the wall, which has likely provided a rigid
support for the mullion. In general, more flexible supports allow more
glazing frame deformation and are less likely to damage the glass. The

2.11 Glass breakage in punched window system during Loma Prieta (magnitude 6.9) Earthquake of October 17, 1989 (Oaklandlibrary.org/oaklandhistory/earthquake89).

2.12 Glass breakage in several storefronts during Loma Prieta (magnitude 6.9) Earthquake of October 17, 1989 (armory.com/~images/?s=quake89).

figures also show that most types of glazing systems discussed in the previous section are indeed vulnerable.

Reports from past earthquakes describe varying levels and different types

2.13 Damage to storefront during Northridge (magnitude 6.7) Earthquake of January 17, 1994 (FEMA News Photo, FEMA, Washington, DC).

2.14 Glass damage to airport control tower during Seattle (Nisqually) Washington (magnitude 6.8) Earthquake of February 28, 2001 (propertyrisk.com/refcentr/seattleeq; photo by ABSG Consulting, 16855 Northchase. Houston, Texas 77060, www.absconsulting.com).

of glazing damage. Following the 1971 San Fernando Earthquake, a survey of 50 high-rise buildings in locations away from the epicenter with only mild shaking and no major structural damage showed that 15 of the surveyed buildings (30%) sustained broken glass (FEMA, 1994). Based on a study by Evans *et al.* (1988), after the 1985 Mexico City Earthquake, glass damage

2.15 Glass breakage in strip window system during Fukuoka (Japan) (magnitude 7.0) Earthquake of March 20, 2005 (answers.com/topic/2005-fukuoka-earthquake).

occurred in over 50% of the buildings that sustained some kind of structural damage. Flexible buildings with structural damage experienced 300% to 400% more glass damage as compared to more rigid buildings that sustained structural damage. A study by Sakamoto *et al.* (1984) showed that larger panels are more susceptible to damage under interstory drift compared to smaller panels. Some of the mentioned examples reflect earthquake events of the 1970s or 1980s when seismic provisions of building codes were still in the development stages; nonetheless, such a review is important since most existing commercial buildings are still clad with the original glazing. Of course, the glazing system in buildings that considered seismic provisions for glazing are expected to show improved performance as compared to older buildings.

Following the Northridge Earthquake of 1994, EERI (1995a) reported in great detail damage to glazing systems, summarized as follows:

- In areas where general nonstructural damage was significant, glazing damage was extensive, and in fact was observed even in areas where other types of nonstructural damage were minimal and rare. For example, downtown Burbank sustained damage to about 25% of the storefront windows in some commercial blocks, even though only 14 buildings in the entire city of population 90 000 were tagged and in general other (than glazing) nonstructural damage was very limited and hardly any structural damage was reported.
- In general, damage to low-rise storefront windows was more extensive than the damage observed in CW systems on high-rise buildings. Reasons for the better behavior of CWs on high-rise buildings include the following: (a) glazing systems on buildings constructed during the prior two decades were designed based on (building code) drift requirements; (b) the glass panes in CWs on older high-rise buildings

2.16 Glass breakage in a storefront in San Jose during San Francisco Bay Area (magnitude 5.6) Earthquake of October 30, 2007 (news. bostonherald.com, Oct. 31, 2007).

were generally smaller than the panes in storefronts and thus sustained less damage; of course, this may no longer be the case, as today's multistory building tenants desire larger glazing, which should perform satisfactory if designed according to the code's drift requirement; (c) satisfactory seismic performance of some CW designs was evaluated through mock-up testing that confirmed sufficiency of glass to frame clearances; and (d) it is speculated that the low-rise storefront glass in older buildings was usually installed without consideration of the requirement for drift capacity or 'rattle space', which means the glass was placed within the glazing frame with small glass-to-frame clearances. Although lack of sufficient clearance is a plausible explanation of the reconnaissance report (EERI, 1995a) for damage to storefront glass, it is possible that other factors, such as the rigidity of the storefront (as compared to horizontal members of curtain walls), may also have influenced such behavior. Full-scale laboratory experiments with realistic boundary conditions are needed to investigate the effects of various factors on storefront behaviors.

- The EERI report (1995a) cites one example on the cumulative damage to glass from aftershocks. It describes an all-glazed entrance to a hotel/ conference center that experienced glass panel rotation and some gasket pullout during the main shock, but later in the day an aftershock with a magnitude of 5 (much weaker than the main 6.7 magnitude) caused complete breakage of the glass at that entrance.
- Buildings with safety film applied to glazing for either protection against burglar, impact, or earthquake damage performed better than uncoated

glazing. However, there were also instances where the glass pane in film-coated windows fell out as a unit like a 'wet blanket', primarily because in such applications the film had been 'unanchored' (also known as 'daylight application'), which means that the film was not connected to the glazing frame to help retain the glass in the event of breakage.

• SSG CWs in general performed better than conventionally glazed systems in earthquakes, but especially better than those dry-glazed systems with 'roll-in' gaskets with no edge blocks, or without sufficient 'edge bite'. These observations are reported in EERI (1995a) based on a survey by the glazing industry.

As widely publicized in the news media at the time, nonstructural damage surrounding glass failure at Sea Tac Airport during the 2001 Nisqually Earthquake was extensive and reduced the critical service at the airport to 50% normal capacity. According to the Nisqually Earthquake preliminary report (EERI, 2001), nonstructural damage can cause business interruption and contribute to economic losses. A 'life-safety' criterion, therefore, is not necessarily sufficient, and a more restrictive 'damage control' criterion may be prudent in a seismic mitigation program. Such distinctions become more important for retrofit decision making as well, as in the performance-based design of new buildings. Currently, the Applied Technology Council (ATC) is developing the Next-Generation Performance-Based Seismic Design Criterion as the ATC-58 project. Under such design criteria, the performance of a building in an earthquake will be evaluated (as part of the design process) in terms of the probability of damage and cost of repair. With this approach, the projected performance outcomes for different structural and nonstructural design alternatives can be compared before a final decision on the choice of the system to use is made.

Current building code requirements for the seismic design of CWs are based on preventing glass fallout, which is a life-safety criterion. However, minimizing cracking can lead to savings of significant repair costs and consequential damage. Minimizing cracking may not be cost effective, as one must balance the cost of minimizing the likelihood of cracking with the probable life-cycle savings in reduced breakage. Such risk/reward analyses should be considered in design decisions.

Glass panels with even one visible crack must be replaced. The question of whether to consider a glass panel with one visible crack a serviceability failure or an ultimate (life-safety) failure warrants an explanation. If glass cracking is such that it can lead to glass fallout (e.g. the existence of several cracks), then it should be considered a failure and should be replaced immediately. On the other hand, for a glass panel with a small crack in a corner still captured by the frame, the question would be the urgency of replacement – not the need for replacement. In general, the strength of a cracked glass panel is significantly diminished, and should be considered a risk for fallout.

Another point relevant to the repair costs is that, in general, unlike an AN monolithic glass pane, laminated glass and IG units may also cause damage to the glazing framing as well because of the inherent higher strength of such configurations. Repair of glazing systems due to earthquake damage can vary greatly and may include glass panel replacement and repositioning. If the glass panel experiences large shifting within the glazing frame it might 'de-glaze' from the framing, causing air infiltration and water leakage, and overstressing of the glass due to loss of support. Subsequent repair work can also affect normal building functions such as occupants' inconvenience, security, and weather damage to building contents. Although glass cracking may not immediately pose life-safety hazards as in glass fallout, it should be considered a potential life-safety hazard, especially for multistory buildings. Glass panels that are weakened by edge damage due to earthquakes and are left in place may lead to fallout at a later time due to 'normal' loads from thermal and/or wind effects.

2.4   Review of laboratory experimental studies

This section summarizes experimental laboratory studies aimed at developing a better understanding of the behavior of architectural glass under seismic loading conditions. Such studies, in general, characterize the performance of different types of glazing systems during frame racking and describe various failure modes under increasing drifts. These studies also help identify aspects of the design that can be modified for improved performance. This section provides a summary of chronological developments in experimental studies on seismic performance of architectural glass, and provides discussion of the more recent studies.

Early experimental studies by Bouwkamp and Meehan (1960) and Bouwkamp (1961) evaluated the effect of in-plane racking displacements on window specimens with different glass sizes and framing systems (steel, aluminum, and wood sash) glazed with putty sealant (both hard putty and soft putty). Glass panels broke when one upper and one lower corner of the framing members exerted pressure on the glass. Bouwkamp (1961) developed simple equations to predict drift that causes glass failure and compared the theoretical results with experimental results. Bouwkamp experimentally studied multiple window panels as well as single panels.

Major studies were carried out 30 years later by New Zealand researchers (Lim and King, 1991; Thurston and King, 1992; Wade, 1990; Wright, 1989) on full-scale racking tests of different curtain wall systems. In-depth studies were carried out shortly thereafter in the US by Behr and his colleagues (Behr, 1998; Behr and Belarbi, 1996; Behr et al., 1995a, 1995b, Pantelides and Behr, 1994) who evaluated the post-breakage behavior of several types of glass used in curtain walls when subjected to dynamic racking movements.

With a substantial data base now developed, Behr worked with NEHRP to develop seismic design provisions for architectural glass that first appeared in the *2000 NEHRP Recommended Provisions for Seismic Regulations for New Buildings and Other Structures* (BSSC, 2001). Behr also worked with the American Architectural Manufacturers Association (AAMA) to develop a test method, *Recommended Dynamic Test Method for Determining the Seismic Drift Causing Glass Fallout from a Wall System*, which was published as AAMA 501.6 in 2001. The 2000 NEHRP seismic design provisions for architectural glass and the AAMA seismic test method were adopted in ASCE 7-02, which was referenced in the *International Building Code 2003* (ICC, 2003) and the *NFPA 5000 Building Code* (NFPA, 2003). Subsequent editions of ASCE 7 and the model building codes have retained these design provisions and the referenced AAMA laboratory test methods for the seismic design of architectural glass components.

As an example of more recent studies, the study of IG unit performance by Memari *et al.* (2003) can be mentioned as representative of studies that characterize seismic racking performance of glazing systems. In this study, which is briefly reviewed here, the seismic resistance of various architectural glass configurations constructed with one laminated glass pane and one monolithic glass pane were evaluated. All IG units consisted of an annealed pane, which was the interior (inboard) pane, and a laminated exterior (outboard) pane with an argon fill and an anodized aluminum spacer between the panes. Several parameters were varied in the laminated pane of each configuration including glass lite thickness, glass type (i.e. AN, HS, and FT), and PVB interlayer thickness. The paper (Memari *et al.*, 2003) reports the average drift values at the first occurrence of glass cracking in each IG unit pane, glass fallout from the monolithic pane, and pullout and fallout of the entire glass unit for each configuration, along with damage to the aluminum framing. The study showed that the use of IG units with a laminated pane had a significant beneficial effect on the seismic service-ability of glass when these IG units, consisting of AN inboard/laminated AN outboard were compared to IG units constructed with two panes of AN monolithic glass or when compared to standard, non-IG single pane laminated AN glass units. For example, test results showed that drift at cracking of IG units with 6 mm ($\frac{1}{4}$ in) AN monolithic glass inner pane with laminated outer pane IG units was about 20% higher than AN monolithic cracking drift in IG units constructed exclusively with 6 mm ($\frac{1}{4}$ in) AN monolithic glass panes. Furthermore, the drift at cracking of 6 mm ($\frac{1}{4}$ in) AN laminated glass panes within laminated/monolithic IG units was about 60% higher than the cracking drifts of single thickness 6 mm ($\frac{1}{4}$ in) AN laminated glass units. The study showed that the use of IG units, such as the ones tested in the study, can lead to serviceable performance at building code (*IBC 2006*) prescribed drift limits.

2.5   Review of analytical studies

Although standardized analysis and design methods for architectural glass wall systems subjected to out-of-plane loads due to wind are relatively well-developed, analogous methods have not been developed for these wall systems subjected to seismic loads. During interstory drift, conventionally constructed CW framing racks cause architectural glass panels to translate and rotate as rigid bodies within the frame. Experimental studies have shown that when the corners of one diagonal of a glass panel make contact with the deformed frame along its shorter diagonal during racking, contact stresses result in glass damage, weakening the glass and rendering it prone to glass fracture and fallout. Thus, in cases where considerations of seismic effects are part of the design criteria, CW manufacturers typically try to satisfy seismic requirements by providing 'adequate' glass-to-frame clearances, as shown in Fig. 2.1. This is done by specifying clearances that exceed a geometry-based prediction for glass-to-frame contact. These clearances (typically 6 mm ($\frac{1}{4}$ in) to 13 mm ($\frac{1}{2}$ in)), however, are often inadequate to accommodate design earthquakes. Interstory drifts of 32 mm ($1\frac{1}{4}$ in) to 38 mm ($1\frac{1}{2}$ in) are not uncommon for a steel moment frame office building in a significant earthquake. The maximum allowable drift for seismic design is based on ASCE 7-05 and is expressed as a percentage of story height, for example 2% for most framed structures, which is equivalent to 89 mm ($3\frac{1}{2}$ in) for a 4420 mm (14.5 ft) story height.

In general, the drift capacity of CWs can be higher than the drift based solely on glass-to-frame clearance, but there is no one-to-one correlation between interstory drift value and required glass-to-frame clearance since several parameters affect such a relationship. These include glazing frame deformation, glazing frame to structure connection flexibility, mechanism of glazing frame capturing of glass panel edges (dry glazed, wet glazed, structural silicone glazed), glass panel size and configuration, and glass type. However, in some cases, drift capacity could be reduced because of the effects of gaskets in dry-glazed systems, which may lead to a glass panel rotation condition that results in early glass-to-frame contact. In other words, the contact of glass corners along a diagonal to the frame is not necessarily at frame diagonal corners (Memari *et al.*, 2006c). The concern about glass-to-frame clearance-based design and accurate interstory drift predictions have led architectural glass wall system manufacturers to conduct full-scale mock-up tests using laboratory test methods such as those developed and published by the American Architectural Manufacturers

Association (AAMA, 2001) to aid in the design of architectural glass wall systems to withstand seismic loads.

The need for analytical modeling has long been recognized, and some attempts have been made in the past to develop methods to predict the response of glass under drift conditions. To date, relatively few published works have addressed the seismic in-plane structural behavior of architectural glass analytically. Bouwkamp (1961) and Bouwkamp and Meehan (1960) pioneered the development of a simple design equation that estimates glass cracking drift capacity in windows. The equation predicts the drift at which the corners along a diagonal of the glass panel will be in contact with the corresponding corners of the window frame when deformed under lateral load. This relation gives the drift required to cause glass-to-frame contact for a given rectangular window panel due to rigid body movement of the glass panel as a function of geometric properties of the glazed panel:

$$\delta r = 2c(1 + h/b) \tag{2.1}$$

In Equation [2.1], δr is the lateral drift required for the glass panel to make contact along both corners of the shorter diagonal of the deformed frame, c is the glass-to-frame clearance (assumed in Equation [2.1] to be uniform around the entire glass perimeter), h is the glass panel height, and b is the glass panel width. A more general form of Equation [2.1] that considers different clearances at horizontal and vertical glass panel edges appears in ASCE 7-05 (ASCE, 2006) as shown below:

$$D_{clear} = 2c_1[1 + (h_p c_2)/(b_p c_1)] \tag{2.2}$$

where D_{clear} is the drift over the height of the glass panel that causes initial glass-to-frame contact, h_p and b_p are, respectively, the height and width of the rectangular glass panel, c_1 is the clearance (gap) between the vertical glass edges and the glazing frame, while c_2 is the clearance between the horizontal glass edges and the glazing frame, as shown in Fig. 2.17.

Such equations can be used by curtain wall designers to estimate the horizontal racking displacement limit (drift), beyond which glass distress would be expected, for a given dry-glazed architectural glass CW system. These equations can also be used to estimate the glass-to-frame tolerances required for a given dry-glazed system to achieve a particular drift capacity without experiencing glass-to-frame contact. These equations assume that sliding and rotation of the glass panel is unrestrained. This, of course, is idealized since there is friction between the glass panel perimeter and the framing/glazing gaskets, setting blocks, or anti-walk pads. Movement of the glass panel is also influenced by the interaction between the corners of the glass panel and the glazing lip of the horizontal frame member that supports the glass (see Fig. 2.18 for a glazing lip definition). Therefore, these

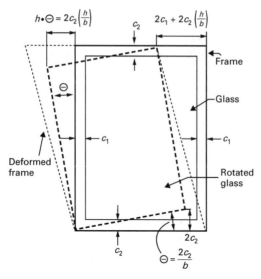

$$h \cdot \ominus = 2c_2\left(\frac{h}{b}\right)$$

$$2c_1 + 2c_2\left(\frac{h}{b}\right)$$

$$\ominus = \frac{2c_2}{b}$$

2.17 Definition of clearances and drift components for use in Equation [2.2].

equations may yield an unrealistic prediction of the onset of glass-to-frame contact and glass damage, as shown subsequently.

As an example, using Equation [2.1], δr is computed as 48 mm (1.90 in) for the glazing system shown in Fig. 2.18. This means that at a drift level of 48 mm (1.90 in) for the 1520 mm (59.84 in) wide by 2060 mm (81.10 in) high frame shown in the figure, glass-to-frame contact is expected to occur. According to the building code (*IBC 2006*), without the availability of experimental data for $\Delta_{fallout}$, defined as the drift that causes a piece of glass at least 1 in^2 in size to fall out of the panel, it may be assumed that the value obtained from the equation is to be the maximum capacity of the glass panel for design purposes, as implied by the first exemption in Section 13.5.9.1 of ASCE 7-05 (2006), which will be discussed further in Section 2.6.

The accuracy of this assumption can best be checked by full-scale testing. Figure 2.19 shows the results of an extensive program of dynamic racking experiments carried out on several different types of glazing system configurations of the size shown in Fig. 2.18 (Behr, 1998). The plotted results show the drift values at the first initial contact between the glass corner and glazing frame, the drift values at the first cracking, and the drift values at $\Delta_{fallout}$. The results shown are average values of the number of tests indicated at the bottom of the figure. Most of the drift causing initial contact results in Fig. 2.18 are around 25 mm (1.00 in), which is significantly different from the 48 mm (1.90 in) value predicted by Equation [2.2]. It should be noted that such a comparison result is for IG unit systems. For monolithic units and lighter glass panes, the result of this comparison is

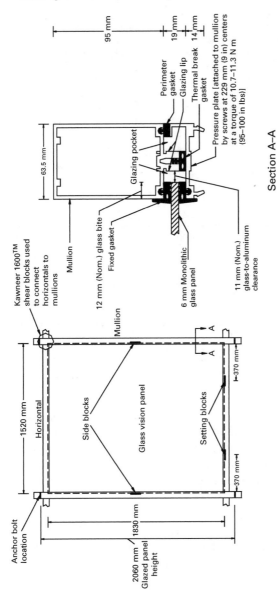

Section A–A

95 mm

19 mm

14 mm

63.5 mm

Perimeter gasket

Glazing lip

Thermal break gasket

Pressure plate [attached to mullion by screws at 229 mm (9 in) centers at a torque of 10.7–11.3 N m (95–100 in lbs)]

Glazing pocket

Mullion

12 mm (Nom.) glass bite

Fixed gasket

6 mm Monolithic glass panel

11 mm (Nom.) glass-to-aluminum clearance

Kawneer 1600™ shear blocks used to connect horizontals to mullions

Mullion

Side blocks

Glass vision panel

Setting blocks

1520 mm

Horizontal

370 mm

370 mm

A

A

Anchor bolt location

2060 mm Glazed panel height

1830 mm

2.18 Detail of dry-glazed curtain wall mock-up used for testing.

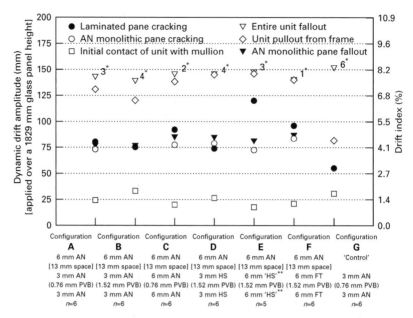

*Indicates the number of specimens that exhibited no glass fallout by the end of the Crescendo Test.
**Configuration E specimens were actually FT (surface compressive pressures of 76.5 MPa (11 100 psi)).

2.19 Racking test results for various insulating glass configurations.

expected to show closer correlation. An alternative to using the exemption based on Equation [2.2] is that $\Delta_{fallout}$ values for a given glazing system should be obtained experimentally or analytically to compare with the value determined from the building structural analysis or code-prescribed allowable drifts. Experimental determination of $\Delta_{fallout}$ values is very costly if it is to be done for each project, and analytical approaches have not yet been developed for this purpose.

Sucuoglu and Vallabhan (1997) presented a glass failure prediction model based on in-plane plate buckling for dry-glazed architectural glass panels under seismic induced in-plane diagonal forces. In this work, drift capacity is obtained by combining the drift from Equation [2.1] and the drift corresponding to diagonal shortening as a result of glass plate buckling. This failure prediction modeling approach is based on the glass plate flexural tensile strength, which must be known to the designer, along with other glass properties such as modulus of elasticity, in order to estimate the drift components due to diagonal buckling of the glass plate. Such a modeling assumption based on flexural stresses has also been used to develop glass thickness charts under wind loading conditions and has been shown to be a valid assumption for out-of-plane bending failure (Beason and Morgan, 1984; Beason et al., 1998).

Although for very large and thin glass plates buckling under in-plane loading could govern the horizontal drift response, full-scale racking tests (e.g. Behr, 1998; Memari et al., 2003, 2004, 2006c) have shown that failure in conventional sizes and thicknesses of glass panels under in-plane racking loading is initiated and governed by local glass edge crushing and subsequent glass cracking at corner regions. Repeated observations during in-plane racking tests on a number of dry-glazed glass and framing types have confirmed that glass damage initiates as a result of glass-to-frame contacts that occur along glass panel edges in corner regions. The flaw distribution at glass edges differs from that on the face of glass panels, and perhaps this difference should be accounted for in glass failure prediction models. On the other hand, it is also possible that the main issue is not that of flaw distribution, but rather the propensity for damage due to glass/metal contact resulting from differing glass edge treatments. In any case, only limited work has been done either experimentally to measure or to model the failure of glass due to edge stresses (Beason and Lingnell, 2002; Carre and Dauderville, 1999; Pantelides et al., 1994), and none of these studies has considered seismic loading along glass edges. Of course, glass strength along the edges is also a function of how the edge is finished (Memari et al., 2006c, Schwartz, 1984a, 1984b). For example, glass panels with cut, ground, seamed, or polished finishes will differ in their crack initiation drift capacity because flaw characteristics of each finish type will differ in their severity and distribution. Edge flaws created by scoring (as when annealed monolithic glass is 'scored and broken' to fabricate given panel dimensions) are points of severe stress concentrations that generally weaken the glass (Schwartz, 1984a, 1984b). This fabrication method may accelerate the glass failure even at normal service loads.

Published studies related to the use of finite element analysis to predict the performance of glazing systems under seismic loads are scarce. One reason for the slow development in this area is that curtain walls are classified as 'nonstructural elements', which implies a lack of justification for the efforts required for advanced structural analysis. Perhaps more importantly, parameters that should be taken into account for modeling are considerable and include, in part, the glazing system materials, the configuration of glazing system and its attachment to the building structural system, the flexibility of the structural frame, the type and thickness of the glass pane and its clearance from the glazing frame, and the material, type, and properties of the gaskets and sealants. Consideration of such parameters makes finite element modeling quite complex. Nonetheless, finite element modeling and analysis may provide an effective means to analyze existing and new curtain wall systems subjected to seismic loading. Of course, it should be noted that this problem may best be expressed in a probabilistic formulation since there is no clear 'allowable stress' for glass under seismic loading conditions.

Nonetheless, as an example of such finite element modeling efforts, Memari *et al.* (2007) developed a finite element modeling approach to predict stresses in the glass panels. The study was conducted to guide the development of a finite element formulation for the analysis of architectural glass curtain walls under in-plane lateral loads. This study is one aspect of ongoing efforts to develop a general prediction model for glass cracking for architectural glass storefront and CW systems during seismic-induced drift. This study was limited to displacement-controlled static loading of a dry-glazed glass CW panel. Physical mock-ups of a dry-glazed glass CW, instrumented with strain gages mounted at select locations on the glass and the aluminum framing (Fig. 2.20) to determine strains when the mock-up was subjected to displacement-controlled loading, helped to improve the finite element modeling (Fig. 2.21) for seismic damage prediction (based on drift capacity) in dry-glazed architectural glass CWs.

This finite element model needs refinement to consider all parameters properly, which include glass panel configuration and glazing frame-to-structure connections. Determination of the stress–strain and load–deformation properties for various components must be obtained through full-scale laboratory testing in follow-up studies.

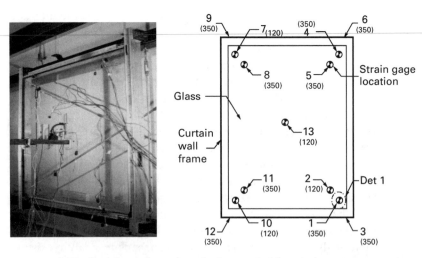

2.20 Curtain wall mock-up instrumented for strain measurement.

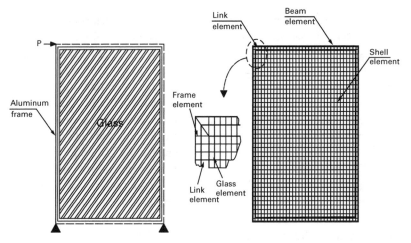

2.21 Finite element model used to predict stresses in the dry-glazed curtain wall mock-up shown in Fig. 2.19.

2.6   Mitigation of seismic damage to glazing systems

The seismic design provisions for glass in glazed systems in ASCE 7-05 are based on a life-safety criterion, which requires the drift causing glass fallout ($\Delta_{fallout}$) to be at least 25% larger than the building design drift multiplied by the building importance factor. The building design drift (ASCE 7-05) is obtained using the calculated elastic seismic loads (which are reduced by the response modification factor R) through elastic analysis and then amplified using the displacement amplification factor (C_d). The rationale behind such a requirement is that the glazing system should accommodate the building story drift without posing a life-safety hazard. Fallout $\Delta_{fallout}$ is defined as the drift that can cause a piece of glass equal to or larger than 645 mm^2 (1 in^2) to break away from the glass panel and fall out, which could pose a life-safety hazard. ASCE 7-05 prescribes engineering analysis or following the procedure described in the AAMA 501.6 test method (AAMA, 2001) to determine $\Delta_{fallout}$ experimentally. Because techniques for modeling and analyzing glazing systems are not yet sufficiently developed to estimate $\Delta_{fallout}$ reliably, mock-up testing is the best available method to demonstrate that the code design criterion for $\Delta_{fallout}$ has been satisfied. As an alternative to satisfying the $\Delta_{fallout}$ design criterion, ASCE 7-05 allows designers to use any of the following three exemptions: (a) providing sufficient glass-to-frame clearance to avoid any glass-to-frame contact; (b) using fully tempered glass for windows no more than 3 m (10 ft) above a walking surface; or (c) using laminated glass installed with perimeter anchorage. Little information is available on the extent of the applications of ASCE 7-05 exemptions.

Aside from ASCE seismic provisions, the Glass Association of North America (GANA, 2004) provides some general guidelines for construction of glazing systems for improved seismic performances. In general, such guidelines intend to provide cushioned supports at the bottom edge and side edges of the glass to maintain an adequate clearance such that the glass lite can be free to move ('float') within the glazing frame without touching the frame. The guidelines include recommendations for the use of setting blocks, side (edge) blocks (also known as anti-walk blocks), and sometimes corner cushions. Setting blocks consist of two elastomeric material supports (e.g. neoprene, EPDM) typically located at quarter points of the sill glazing member. Side blocks are usually positioned within the top and bottom third of each vertical side. The guidelines also recommend use of cushions at all four corners to prevent glass-to-frame contact. Such cushions help avoid 'hard' contact points but can still load the glass edge/corner and may cause fracture. Glaziers now use 'W-blocks' as side blocks that buckle under significant compressive load so that 'hard' points are avoided until the glass is very close to the metal. Figure 2.22 shows typical locations for setting blocks, side blocks, and corner cushions according to GANA (2004).

Some of the other approaches that have been used in the past to improve the seismic performance of architectural glass, or at least reduce the risks associated with its failure, include (recommended by FEMA 74, 1994) the use of tempered glass, laminated glass, or PET film adhered to the inside glass surface. The first two solutions are applicable to new designs, while the use of PET film is primarily for retrofit of existing glass. The use of *anchored* adhesive films (the film must be anchored to the frame perimeter to be

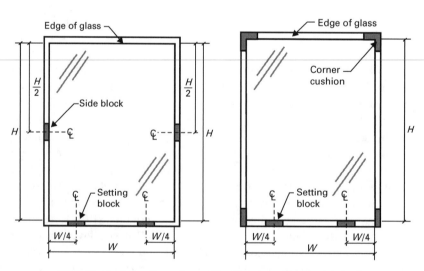

2.22 Typical locations for setting blocks, side blocks, and corner cushions.

effective in this capacity) has been recognized as an effective way to mitigate life-safety hazards due to earthquakes. The use of tempered glass can improve the serviceability drift (i.e. drift level that causes cracking) as compared to annealed glass. On the other hand, the use of laminated glass (composed of annealed glass panes) does not substantially improve the serviceability drift compared to annealed monolithic glass, but it does typically improve the ultimate drift (i.e. drift that causes glass fallout).

Recently, there have been developments in methods for seismic damage mitigation of glazing systems. Seismically isolated CWs (Bruggeman *et al.*, 2000) and unitized systems (Zarghamee *et al.*, 1995) have been used to accommodate earthquake-induced interstory drifts. The basic concept behind such approaches is to isolate the CW frame from the structural frame displacements by modifying the method of attachment of the CW system to the building structure. Another approach that was also recently developed at Penn State is to employ modified geometry at glass corners (rounded corner glass (RCG)) instead of the conventional rectangular corners. Laboratory studies of RCG architectural glass panels have shown that such geometry modification can increase the serviceability drift without the need to change the glazing frame details and its connection to the structural frame. In the following subsections, some of the recent studies on the use of PET film on glass, the RCG concept, and seismic evaluation of SSG systems are reviewed in greater detail (compared to the literature reviewed up to this point) in order to provide the important results of the studies that can be helpful in practical seismic design of glazing systems.

2.6.1 Review of research on the use of safety film

After the 1989 Loma Prieta Earthquake (EERI, 1990) and the 1994 Northridge Earthquake (EERI, 1995a), the potential for filmed glass to mitigate potential injuries as a result of earthquakes was recognized. The importance of this additional intended use for applied film is clear when one realizes that earthquake-induced damage to architectural glass components often necessitates expensive repairs, exposes building contents to weather, theft, and vandalism, causes a disruption of activities within the building, and presents a threat to life safety when glass falls from a wall system. Today, architectural glass fitted with anchored applied film is being used for seismic retrofit applications due to its ability to hold broken glass shards in the glazed opening when the glass is broken (Beason and Lingnell, 2002; Wang *et al.*, 1992). Limited field surveys (EERI, 1995a) have documented the seismic resistance of architectural glass installation with applied film. However, some have reported that glass with protective film (mostly unanchored) has fallen out (Gates and McGavin, 1998).

In recent years, manufacturers have been producing 'safety and security' films to address the need for mitigating damage and injuries caused by flying glass during windstorms, blast loadings, and earthquakes. Prior to the use of films as a retention device, window films were typically applied to the inside face of glass panels in such a way that film was trimmed short of the edges of the glass; i.e. the film was unanchored to the window frame. This form of application, which is referred to as 'daylight application', creates a vulnerable perimeter boundary such that the entire pane of filmed glass can fall out as a 'wet blanket'. To retain broken glass in a reasonably reliable manner, the film must be anchored to the perimeter frame and the frame must resist post-breakage loads. Although the details of anchoring systems vary, they usually involve: (1) the use of an adhesive around the perimeter of the filmed glass to connect the film boundary to the frame; (2) the extension of the film into or on to the perimeter frame and use of a mechanical anchor; or (3) some combination of the two.

A few studies have also been conducted to characterize the seismic performance of architectural glass panels with daylight-applied film (Behr, 1998; Pantelides and Behr, 1994; Sakamoto et al., 1984; Wang et al., 1992). These studies documented the serviceability (glass cracking) and post-breakage behavior of annealed monolithic glass with 0.1 mm (4 mil) PET film during racking mock-up tests which simulated seismic-induced movements of the wall system. Drift time histories similar to that now employed by AAMA (2001) were used in the studies conducted by Pantelides and Behr (1994) and Behr (1998). Although the mock-ups tested in these studies demonstrated the ability of daylight-applied film to hold broken glass shards in the window frame at drift amplitudes that would lead to glass fallout in similarly glazed unfilmed glass specimens, these tests also underscored the vulnerability of glass fitted with daylight applied film to the wet blanket fallout of the entire filmed glass unit.

Memari et al. (2004) studied the seismic performance of architectural glass fitted with anchored applied film. The primary objective of this study was to investigate the response of annealed monolithic architectural glass. The study of the behavior of various applied film anchor systems on annealed monolithic glass panels subjected to cyclic racking tests proved that the anchorage type can influence both the serviceability (glass cracking) and the ultimate (glass fallout) limit states. Three common film-to-frame anchoring methods were evaluated: (1) structural silicone adhesive along the entire glass panel perimeter; (2) an aluminum bar extrusion to anchor the film to the frame horizontal along only the top of the glass panel; and (3) two aluminum bar extrusions to anchor the film to the frame verticals along the two vertical edges of the glass panels. The cyclic racking drift data for the three anchor systems evaluated in this study showed that annealed monolithic glass panels with anchored applied film according to the first, second, and third methods of anchoring can result in glass cracking of the following respective drifts: 66 mm, 45 mm, and 35 mm. Application of film will not increase the capacity of glass against cracking. However, it will increase the fallout drift capacity. In fact, since the glass will be adhered to the film even after the entire panel is cracked, pieces of glass will not easily fall out; rather, the entire unit can fall out as a 'wet blanket' if the film is not anchored to the glazing frame. These results are for 'new' window film installations in one frame type employing a particular glass-to-frame clearance and one glass panel aspect ratio, and must be interpreted within that context.

2.6.2   Review of research on modified corner geometry glass

While most published experimental studies have endeavored to characterize the seismic performance of architectural glass glazing systems as they are conventionally constructed, a few studies have also focused on methods to improve the performance of architectural glass during earthquakes. Each method has its limitations. For example, most seismically isolated wall systems designed to resist earthquakes are tailored primarily for new building construction, not building retrofits, and they tend to be more expensive than conventional wall systems that are not specifically designed for earthquake resistance. As another example, use of anchored safety films and laminated glass increases glass fallout resistance, but does not necessarily increase glass cracking resistance. Furthermore, some earthquake-resistant wall systems limit aesthetic choices in the architectural design of a building's exterior, such as wide mullion wall systems designed for increased clearance to help prevent glass-to-frame contact during earthquake-induced building motions.

In an attempt to develop a simple method intended to reduce glass cracking during earthquakes, Memari et al. (2006c) developed the concept of rounded corner glass (RCG) panels. The in-plane deformation of the glazing frame due to lateral building movements causes the glass panel to translate and rotate as a rigid body within the frame's glazing pocket. Observations made during previous studies showed that square corners and edge protrusions (i.e. flares), which are sometimes present along the edges of glass panels (and especially at corner regions as glass lites are scored and broken to the desired size), tend to act as stress concentration points during glass-to-frame contacts. These observations led to the idea that material removal by rounding the sharp, square glass corners, combined with appropriate glass edge finishing treatment, should result in a higher drift capacity for architectural glass panels because the panels would be able to slide more freely at the corners and reduce or eliminate glass-to-frame corner contacts. Figure 2.23 shows a stock of glass panes with conventional square corners and the RCG concept.

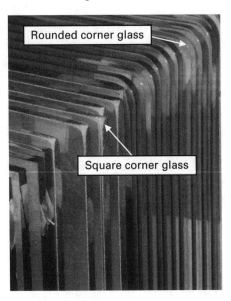

2.23 A stock of squared corner glass and rounded corner glass panes.

Cyclic racking tests revealed that modifying conventional square cornered glass lites by rounding the corners and appropriately finishing the edges (e.g. seamed edge or flat polished edge) increases the glass cracking and glass fallout resistances of those panels within a conventional, dry-glazed wall system. For AN monolithic glass panels, a 13 mm ($\frac{1}{2}$ in) corner radius produced the highest average glass cracking drift and glass fallout drift of all the radii tested in this study (which ranged from 13 mm ($\frac{1}{2}$ in) radius to 76 mm (3.0 in) radius). Thus, with such a small radius at the rounded corners, no modifications to most glazing frame details (e.g. frame width or glazing pocket depth) would be required to hide the corner arc on the glass. Various combinations of edge finish and corner rounding were found to produce enhanced drift limits compared to corner rounding alone. For example, for AN monolithic glass panels, the combination of both a 25 mm (1 in) rounded corner and seamed edges resulted in a 90% increase in the glass cracking drift limit and a 76% increase in the glass fallout drift limit as compared to square cornered AN monolithic glass panels with scored and cut edges. As another example, for FT monolithic glass panels with 25 mm (1.0 in) radius rounded corners and flat polished edges/corners, 50% increases in glass cracking and glass fallout drift limits were observed as compared to the square cornered counterpart with seamed edges. The high drift indices associated with glass cracking and glass fallout for the FT monolithic glass panels with rounded corners and flat polished edges suggests that these configurations could offer serviceable performance even during interstory drifts that are representative of severe earthquakes.

The RCG study also underscored the sensitivity of both square corner and rounded corner FT monolithic glass panels to flares along panel edges in the corner regions (which could be created during the process of removing material from the corners before heat treatment). It should be noted that removing material from the corner does not automatically eliminate flares, and rough edge contours and finish conditions, particularly in the corner regions, can reduce the potential seismic performance advantages of the RCG concept. Due to the simplicity of the RCG concept, properly manufactured architectural glass panels with rounded corners offer an effective glazing option that can help reduce glass damage due to seismic movements in both retrofit applications and in new building construction.

2.6.3 Review of research on structural silicone glazing

Since its introduction more than four decades ago, structural silicone glazing (SSG) systems have become a popular glazing method for CW construction. One important reason for SSG popularity is the aesthetic possibilities associated with an apparently 'mullionless' curtain wall system (Parise and Spindle, 1991; Stubbs, 1986). SSG systems differ from conventionally captured systems in that the glass lites or panels are adhered to the supporting frame with structural silicone sealant along either two edges of the glass (two-side structural support) or all four edges (four-side structural support) (AAMA, 1985). Since SSG systems rely upon the structural silicone sealant to transfer lateral loads to the supporting frame, the strength and modulus properties of the structural silicone sealant and the quality of the bond between the sealant and its substrates (i.e. glass and frame) are of great importance.

There are very few studies reported in the open literature pertaining to the seismic performance of SSG systems (Behr, 1998, Zarghamee et al., 1995). Designers generally assume that SSG systems perform well in seismic regions due to the 'resiliency' of the silicone attachment of the glass to the glazing frame (e.g. ASTM Standard C 1401, 2002). To develop a better understanding of seismic behavior of SSG systems, a study was carried out at Penn State University (Memari et al., 2006a, 2006b) to evaluate the serviceability and ultimate limit state performance of two-sided SSG systems for different glass types and configurations under cyclic racking displacements.

Trial kinematic-based analytical models were also developed to predict failure states (e.g. structural sealant failure) of the two-sided SSG curtain walls with structural sealant on the two vertical edges of the glass pane and rubber gasket captured (i.e. dry glazed) top and bottom edges. Simplifying assumptions were made for model development including the relation of the shear deformation of a two-sided SSG sealant joint to the two-sided SSG

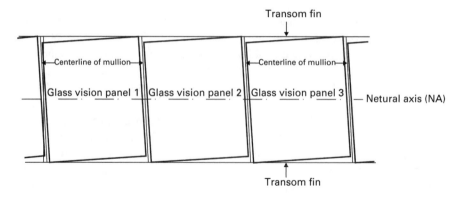

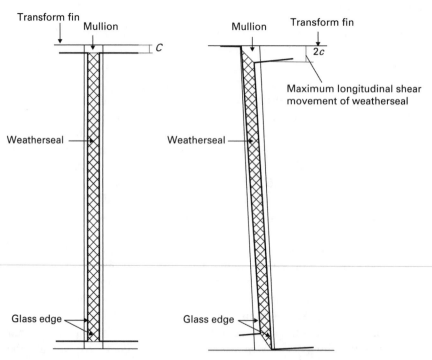

2.24 Idealized movement of two-sided SSG CW frame system and weatherseal silicone deformation.

curtain wall geometry, i.e. a function of glass-to-frame clearance, glass panel width, and glass panel height. In other words, it was assumed that the model estimates the drift capacity by relating it to the shear deformation capacity of structural silicone along the vertical edges of the two-sided SSG. The model was evaluated by using the information from mock-up tests in an experimental part of the study. Figure 2.24 shows the idealized movement of weatherseal at the joint between adjacent vertical edges of glass panel. A corresponding deformation form was also assumed for the structural seal. As an example, the model predicted a drift capacity (without any safety factor) of 83 mm (3.27 in) based on structural seal failure (the model for weatherseal did not show failure of weatherseal), while the experimental results showed structural seal failure in the range of 76 mm (3.00 in) to 102 mm (4.00 in). This shows a good correlation between the model and test results. As for glass cracking drift capacity, the specimen with the lowest structural seal failure drift (76 mm (3.00 in)) showed an experimentally obtained cracking drift capacity of 83 mm (3.25 in) (Memari *et al.*, 2006a). Compared to the experimentally obtained cracking drift capacity of conventionally glazed AN monolithic glass of 39 mm (1.54 in) (Memari *et al.*, 2006c) the SSG specimen shows over 100% higher capacity. This result indicates that if relevant parameters in both types of configurations are the same (e.g. panel dimensions, glazing frame-to-structure frame connection, glass type, loading protocol, etc.), then an SSG configuration of curtain walls will likely result in an increase in drift capacity compared to a conventionally glazed system. It should be noted, however, that comparison should also be made between the performance of the two systems for in-service (aged) conditions as well. Additional research is required to enable such comparsions.

2.7   Future trends and conclusions

Although considerable experimental laboratory research has been under-taken for some curtain wall and storefront systems, there are several types of glazing systems and glass for which the experimental-based knowledge is minimal. Besides experimental mock-up testing to determine drift capacity, ASCE 7-05 permits the use of 'advanced analysis' techniques in the seismic design process of CWs, but experimentally validated analytical modeling approaches for this purpose are not yet well developed. As a result, CW design professionals do not generally employ computer modeling and analysis for CW design. Development of advanced empirically based design and analytical modeling approaches, including finite element modeling to predict the seismic response of CW systems, will help designers gain better understanding of their designs without the need to over-rely on laboratory mock-up test results. In fact, once a validated model is available, it can be analyzed for different loading conditions. Such modeling techniques can also be used in seismic evaluation of existing curtain walls for assessment of retrofit needs. Lack of well-developed seismic design and analysis procedures for curtain walls sometimes leaves design professionals and manufacturers in difficult situations because they cannot present adequate seismic design calculations or analysis results to support their curtain wall design decisions, so they are left with no other way than to resort to mock-up testing.

Development of analytical approaches for seismic assessment and design of CW systems will also allow performance-based designs to be evaluated consistent with other nonstructural components in the design process. Performance-based design processes will require prediction of an array of performance objectives. Predicting the seismic behavior of a given CW system under different levels of seismic input will help designers to choose the most suitable systems. Furthermore, advancement in seismic design methodology for architectural glazing systems will enable the design and development of cost-effective new glass curtain wall systems and help decision making for retrofit of existing ones.

Traditionally, architects conceptualized CW systems, while curtain wall manufacturers engineered, specified, and constructed these systems. Recent building code seismic design requirements for glazing systems have increased the need for architectural engineers and structural engineers to be increasingly involved in the design process for glazing systems. Development of advanced analysis and design tools will further bring engineers into the design process and will likely reduce over reliance on expensive and time-consuming mock-up testing.

2.8   References

American Architectural Manufacturers Association (AAMA) (1985) *Structural Sealant Glazing Systems – A Design Guide*, AAMA Aluminum Curtain Wall Series No. 13, American Architectural Manufacturers Association, Des Plaines, IL.

American Architectural Manufacturers Association (AAMA) (2001) *Recommended Dynamic Test Method for Determining the Seismic Drift Causing Glass Fallout from a Wall System*, Publication No. AAMA 501.6-01, American Architectural Manufacturers Association, Des Plaines, IL.

American Society of Civil Engineers (ASCE) (2006) *Minimum Design Loads for Buildings and Other Structures*, ASCE 7-05, ASCE, Reston, VA.

American Society of Testing Materials (ASTM) (2002) *Standard Guide for Structural Sealant Glazing*, C 1401-02, ASTM International, West Conshohocken, PA.

American Society of Testing Materials (ASTM) (2004a) *Standard Practice for Determining Load Resistance of Glass in Buildings*, E 1300-04, ASTM International, West Conshohocken, PA.

American Society of Testing Materials (ASTM) (2004b) *Standard Specification for Heat-treated Flat Glass – Kind HS, Kind FT, Coated and Uncoated Glass*, C 1048-04, ASTM International, West Conshohocken, PA.

Beason, W. L. and Lingnell, A. W. (2002) A thermal stress evaluation procedure for monolithic annealed glass, *Use of Glass in Buildings*, ASTM STP 1434, V. Block (ed.), ASTM International, West Conshohocken, PA.

Beason, W. L. and Morgan, J. R. (1984) Glass failure prediction model, *Journal of Structural Engineering*, ASCE, 110(2), 197–212.

Beason, W. L., Kohutek, T. L. and Bracci, J. M. (1998) Basis for ASTM E 1300 annealed glass thickness selection charts, *Journal of Structural Engineering*, ASCE, 124(2), 215–221.

Behr, R. A. (1998) Seismic performance of architectural glass in mid-rise curtain wall, *Journal of Architectural Engineering*, ASCE, 4(3), 94–98.

Behr, R. A. and Belarbi, A. (1996) Seismic test methods for architectural glazing systems, *Earthquake Spectra*, 12(1), 129–143.

Behr, R. A., Belarbi, A. and Brown, A. T. (1995a) Seismic performance of architectural glass in a storefront wall system, *Earthquake Spectra*, 11(3), 367–391.

Behr, R. A., Belarbi, A. and Culp, J. H. (1995b) Dynamic racking tests of curtain wall glass elements with in-plane and out-of-plane motions, *Earthquake Engineering and Structural Dynamics*, 24, 1–14.

Bell, G. R. and Zarghamee, M. S. (2004) Impact of recent code changes on the seismic design of building cladding, in Proceedings of the ASCE Structures Congress, May 2004.

Bouwkamp, J. G. (1961) Behavior of window panels under in-plane forces, *Bulletin of the Seismological Society of America*, 51(1), 85–109.

Bouwkamp, J. G. and Meehan, J. F. (1960) Drift limitations imposed by glass, in Proceedings of the Second World Conference on *Earthquake Engineering*, Tokyo, Japan, pp. 1763–1778.

Brueggeman, J. L., Behr, R. A., Wulfert, H., Memari, A. M. and Kremer, P. A. (2000) Dynamic racking performance of an earthquake-isolated curtain wall system, *Earthquake Spectra*, 16(4), 735–756.

Building Seismic Safety Council (BSSC) (2001) *2000 NEHRP Recommended Provisions for Seismic Regulations for New Buildings and Other Structures, Part 1 – Provisions*, prepared by the Building Seismic Safety Council, Washington, DC, for the Federal Emergency Management Agency and issued as FEMA 368.

Carre, H. and Dauderville, L. (1999) Load-bearing capacity of tempered structural glass, *Journal of Engineering Mechanics*, 125(8), 914–921.

Deschenes, J. P., Behr, R. A., Pantelides, C. P. and Minor, J. E. (1991) Dynamic racking performance of curtain wall elements, Technical Report prepared for Monsanto Chemical Company, Saflex Division, and Robertson/Cupples Company of St Louis, MO, NTIS accession No. PB92 - 140631.

EERI (1990). Loma Prieta Earthquake Reconnaissance Report, *Earthquake Spectra*, Supplement to Volume 6, Earthquake Engineering Research Institute, Oakland, CA.

EERI (1995a) Northridge Earthquake Reconnaissance Report, Volume 1, *Earthquake Spectra*, Supplement C to Volume 11, Earthquake Engineering Research Institute, Oakland, CA.

EERI (1995b) The Hyogo-Ken Nanbu Earthquake January 17, 1995 Preliminary Reconnaissance Report, EERI 95-04, Earthquake Engineering Research Institute, Oakland, CA.

EERI (2001) The Nisqually, Washington, Earthquake February 28, 2001

Preliminary Reconnaissance Report, EERI 2001-01, Earthquake Engineering Research Institute, Oakland, CA.

Evans, D., Kennett, E., Holmes, W. T. and Ramirez, F. J. L. (1988) Glass damage in the September 19, 1985 Mexico City Earthquake, Report prepared for NSF, CES-861093.

FEMA (1994) *Reducing the Risk of Nonstructural Earthquake Damage – A Practical Guide*, FEMA 74, Federal Emergency Management Agency, Washington, DC.

FGMAJ (1995) Glass Damage Report – A report (explanation) on damage to window glass in the Great Hanshin Earthquake, Flat Glass Manufacturers Association of Japan, Translation from Japanese.

GANA (2004) *Glazing Manual 2004 Edition*, Glass Association of North America, Topeka, Kansas.

Gates, W. E. and McGavin, G. (1998) Lessons learned from the 1994 Northridge Earthquake on the vulnerability of nonstructural systems, in Proceedings of the Seminar on *Seismic Design, Retrofit, and Performance of Nonstructural Components*, ATC 29-1, Applied Technology Council, pp. 93–106.

International Code Council (ICC) (2003) *International Building Code (IBC) 2003*, ICC, Falls Church, VA.

International Code Council (ICC) (2006) *International Building Code (IBC) 2006*, ICC, Falls Church, VA.

King, A. B. and Thurston, S. J. (1992) The racking behaviour of wall glazing during simulated earthquake, in Tenth World Conference on *Earthquake Engineering*, Volume 6, Madrid, Spain, July 1992.

LaTona, W., Schwartz, T. A. and Bell, G. R. (1988) New standard permits more realistic curtain wall testing, *Building Design and Construction*, November 1988, pp. 42–46.

Lim, K. Y. S. and King, A. B. (1991) The behavior of external glazing systems under seismic in-plane racking, Building Research Association of New Zealand (BRANZ), Study Report No. 39.

Lingnell, A. W. (1994) Initial survey and audit of glass and glazing system performance during the earthquake in the Los Angeles Area on January 17, 1994, Final Report submitted to Primary Glass Manufacturers Council, Lingnell Consulting Services.

Memari, A. M., Behr, R. A. and Kremer, P. A. (2003) Seismic behavior of curtain walls containing insulating glass units, *Journal of Architectural Engineering*, ASCE, 9(2), 70–85.

Memari, A. M., Kremer, P. A. and Behr, R. A. (2004) Dynamic racking crescendo tests on architectural glass fitted with anchored 'PET' Film, *Journal of Architectural Engineering*, ASCE, 10(1), 5–14.

Memari, A. M., Chen, X., Kremer, P. A. and Behr, R. A. (2006a) Seismic performance of two-side structural silicone glazing systems, *Journal of ASTM International (JIA)*, 3(10), 1–10.

Memari, A. M., Chen, X., Kremer, P. A. and Behr, R. A., (2006b). Development of failure prediction models for structural sealant glazing systems under cyclic racking displacement conditions, in Proceedings of 2006 Architectural Engineering Conference on *Building Integration Solutions*, Omaha, Nebraska, March 30–April 2, 2006, CD-ROM, 15 pages.

Memari, A. M., Kremer, P. A. and Behr, R. A. (2006c). Architectural glass panels with rounded corners to mitigate earthquake damage, *Earthquake Spectra Journal*, 22(1), 129–150.

Memari, A. M., Shirazi, A. and Kremer, P. A. (2007) Static finite element analysis of architectural glass curtain walls under in-plane loads and corresponding full-scale test, *Structural Engineering and Mechanics Journal*, 25(4), 365–382.

National Fire Protection Association (NFPA) (2003) *Building Construction and Safety Code^{TM} NFPA 5000*, National Fire Protection Association, Quincy, MA.

Pantelides, C. P. and Behr, R. A. (1994) In-plane racking tests of curtain wall glass elements, *Earthquake Engineering and Structural Dynamics*, 23, 211–228.

Pantelides, C.P., Sallee, G.P. and Minor, J.E. (1994) Edge strength of window glass by mechanical test, *Journal of Engineering Mechanics*, 120(5), 1076–1090.

Parise, C. J. and Spindler, R. G. (1991) Structural sealant glazing in the 1980s, *Exterior Wall Systems: Glass and Concrete Technology, Design, and Construction*, ASTM STP 1034, B. Donaldson (ed.), American Society for Testing and Materials, West Conshohocken, PA, pp. 94–117.

Sakamoto, I., Itoh, H. and Ohashi, Y. (1984) Proposals for seismic design method on nonstructural elements, in Proceedings of the 8th World Conference on *Earthquake Engineering*, Volume 5, San Francisco, CA, pp. 1093–1100.

Schwartz, T. A. (1984a) How to avoid glass fracture, *Glass Digest*, March 15, 1984, 58–61.

Schwartz, T. A. (1984b) Analyzing the cause of fracture after the fact, *Glass Digest*, April 15, 1984, 66–74.

SFGATE (2007) http://www.sfgate.com/cgi-bin/object/article?f=/c/a/2007/07/20/BAquake.DTL&o=1

Stubbs, M. S. (1986) Glued-on glass – examining structural silicone sealant glazing, *Architectural Technology*, May/June 1986, 46–51.

Sucuoglu, H. and Vallabhan, C. V. G. (1997) Behavior of window glass panels during earthquakes, *Engineering Structures*, 19(8), 685–694.

Thurston, S. J. and King, A. B. (1992). Two-directional cyclic racking of corner curtain wall glass elements, Building Research Association of New Zealand (BRANZ), Study Report No. 44.

Wade, C. A. (1990) Structural performance of conservatories, Building Research Association of New Zealand (BRANZ), Study Report No. 30.

Wang, M. L., Sakamoto, I. and Bassler, B. L. (1992) Design of cladding for earthquakes, Chapter 4, in *Cladding*, Council on Tall Buildings and Urban Habitat, McGraw-Hill, Inc., New York, pp. 82–84.

Wright, P. D. (1989) The development of a procedure and rig for testing the racking resistance of curtain wall glazing, Building Research Association of New Zealand (BRANZ), Study Report No. 17.

Zarghamee, M. S., Schwartz, T. A. and Gladstone, M. (1995) Seismic behavior of structural silicone glazing, in *Science and Technology of Building Seals, Sealants, Glazing and Waterproofing*, Volume 6, ASTM STP 1286, James C. Myers (ed.), American Society for Testing and Materials, West Conshohocken, PA, pp. 46–59.

3

Snow loads on building envelopes and glazing systems

R. FLOOD, Matrix IMA, USA

Abstract: This chapter discusses snow and snow/ice issues on building envelopes with a particular emphasis on architectural glazing. Topics include snow load data sources, the conversion of ground snow load to basic flat roof design loads, and subsequently to sloped roof design loads specifically related in application to skylights and sloped glazing. The conversions are based on the International Building Code mandated use of the engineering data contained in ASCE 7, Chapter 7, 'Snow Loads'. Other roof snow glazing issues are discussed as well as vertical glazing snow hazards.

Key words: ground snow load, flat roof snow load, sloped roof snow load.

3.1 Introduction

The intent of this chapter is not to select the glazing type, thickness, strength, size, or durability needed to withstand extreme snow/ice conditions. The intent is to help the design professional identify and quantify the realistic extreme snow/ice forces and dangers acting on the building envelope, especially architectural glazing.

This chapter deals with snow loads and snow/ice issues related to the building envelope with particular emphasis on sloped glass/skylights. The first section discusses data sources for obtaining snow loads, primarily focused on the United States. The next section discusses roof snow load design methodology used in the United States per the US Standard ASCE 7, with a focus on sloped glazing/skylights. The third section discusses other roof snow glazing issues such as snow bridging, roof snow retention, ice dam/icicle and sliding snow/ice dangers, and melt water drainage. The last section discusses snow issues for vertical glazing.

3.2   Snow load sources

3.2.1 Codes

The 2006 International Building Code (IBC 2006)[1] and *2006 International Residential Code* (IRC 2006)[2]

The IBC 2006 Section 1608, 'Snow Loads', requires that design snow load for the building 'shall be determined in accordance with Chapter 7 of ASCE 7'. This chapter of the American Society of Civil Engineers (ASCE) Standard 7 will be discussed in section 3.2.2. The IBC and IRC further require that the ground snow load used to determine the design snow load also be per ASCE Standard 7-05 or per the map in IBC Figure 1608.2 (Figure R301.2 (5) in IRC). The ground snow load maps in the IBC, IRC, and ASCE 7 are identical (see Fig. 3.1). The map was made by Wayne Tobiasson and Alan Greatorex of the Cold Regions Research and Engineering Laboratory (CRREL) and first published in the 1995 edition of ASCE 7. As the topic of this publication is about glass resisting extreme climatic events, the snow load determination will be per ASCE 7.

3.2.2 Standards

The ASCE 7-05 Minimum Design Loads for Buildings and Other Structures (ASCE 7-05)[3]

Chapter 7 of ASCE 7-05 deals with snow loads and Section 7.2 discusses ground snow loads and references ASCE 7-05 Figure 7-1, which is the ground snow load map of the contiguous United States (Fig. 3.1). The map background indicates states and counties and the heavy solid contour lines indicate different ground snow load zones. The number within a zone is the ground snow load in lb/ft^2 ($0.0479 \ kN/m^2$). Numbers in parenthesis above these loads indicate the upper elevation in feet (0.3048 m) for the ground snow load value presented below. In some zones several loads and elevation limits are provided. At higher elevations a case study (CS) analysis must be made to determine the ground snow load. In other zones, where extreme local variations precluded mapping, the entire zone is labeled 'CS' to indicate that case study analyses are required for all elevations therein.

A 'CS' analysis is for a site-specific location, i.e. exactly where your project is located geographically by latitude and longitude and elevation (height) above mean sea level. The analysis also must be made by an extreme value statistical analysis of ground snow load data available in the vicinity of the site using values having a 2 % annual probability of being exceeded. Such values are also known as 50 year mean recurrence interval (50 year

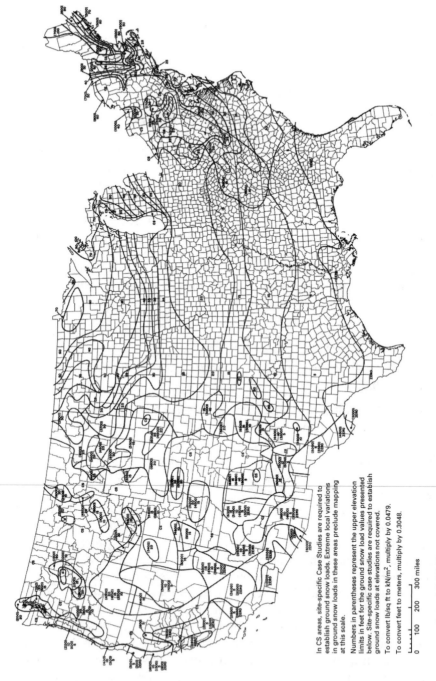

In CS areas, site-specific Case Studies are required to establish ground snow loads. Extreme local variations in ground snow loads in these areas preclude mapping at this scale.

Numbers in parentheses represent the upper elevation limits in feet for the ground snow load values presented below. Site-specific case studies are required to establish ground snow loads at elevations not covered.

To convert lb/sq ft to kN/m², multiply by 0.0479.

To convert feet to meters, multiply by 0.3048.

0   100   200   300 miles

3.1 ASCE Figure 7-1 – reprinted with permission from ASCE.

MRI) values. Some Alaskan locations are given in ASCE 7-05 Table 7-1. For other Alaskan locations case study analyses are required.

A thorough discussion of the ASCE 7-05 ground snow provisions is in the Commentary Chapter, C7, Section C7.2 'Ground Snow Loads, P_g' . Table C7-2 shows high elevation case study loads that are above the altitude limits of Figure 7-1 map zones and Table C7-3 provides factors for converting different MRI values to the 50 year MRI standard.

Lastly, ASCE 7-05 Commentary Chapter C7 'References' C7-1 to C7-15 (C7-16 Canada) provide additional ground snow data. Researching these referenced data for a given locale should help in determining the appropriate ground snow load for design purposes.

3.2.3 Local jurisdictional authority

The local building official (or other entity charged with enforcing building regulations) can be a source for appropriate snow load information within their area of jurisdiction, but not always. The wise professional will obtain a written/published snow load value from the building official, but will always cross check such a value with information that meets the requirements of ASCE 7-05. Be sure to determine if the furnished value is a ground snow load or a design 'flat roof snow load'. In some instances, it may include sloped roof loads up to certain slopes. Ground snow load (P_g) is the snow weight on the ground. Flat roof snow load (P_f) is the snow weight on a flat roof, the starting point for roof design snow loading.

This is an important distinction since the 'flat roof snow load' is usually about 30% less than the ground snow load. The basic fixed '0.7' conversion factor of ground snow load to flat roof snow represents a conservative average derived from the O'Rourke et al. (1983)[4] CRREL study. However, in some instances the local jurisdiction has not accounted for the other multiplying factors in the ground snow load to flat roof snow load conversion equation (ASCE 7-05 Equation 7-1). Such misunderstandings have caused both underdesign and overdesign of the roof, especially the sloped glazing/skylights.

In some jurisdictions, where case studies are required, regionally produced data are used to establish either ground snow load or 'design' snow load (assumed to be flat roof snow load, but should be verified with the local jurisdiction). These regional data are sometimes expressed graphically as curves such as shown in Fig. 3.2. These curves correlate snow load with altitude. The local jurisdiction would identify which curve to use.

Full reliance on this snow load versus altitude data curve may be misleading as the data do not take into consideration either site location or wind factors. As an example, a house part way up a south-facing hill on the

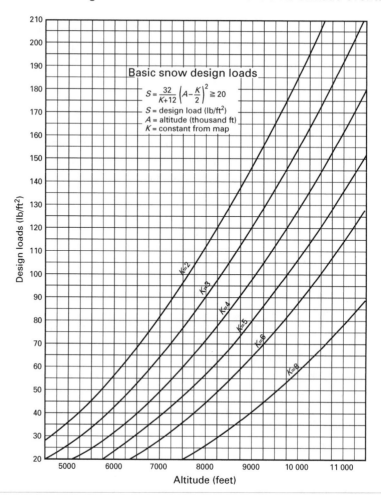

3.2 Regional snow load versus altitude curves. Local jurisdictional authority would select appropriate curve (Matrix IMA Archive: public record document).

north side of a wide valley will probably have less snow than the same house at the same elevation located on the south side of the valley in the shadow of a mountain range. If the storm and wind track is from the south/southwest, the south valley house will also be on the lee (north) side of the mountain. Lack of winter sun combined with minimal wind stripping can produce much higher snow loads than derived from data such as Fig. 3.2. The minimum snow load from these curves should be adjusted for the specific project site in accordance with the multiplying factors in ASCE 7-05 Equation 7-1. An example is presented in Section 3.2.5.

3.2.4 Site-specific case study

The ground snow load maps in the IBC, IRC, and ASCE 7-05 standards all indicate that 'CS' areas require a 'site-specific case study' snow load analysis. Tobiasson and Greatorex[5] developed a process for determining the ground snow load at any given site within the continental United States and Alaska. For the specific site in question, the elevation above mean sea level and its latitude and longitude coordinates are required. To paraphrase O'Rourke[6] the process involves regressing the known 50 year ground snow load values versus elevation from a number of nearby recording stations. The recording station loads are plotted and a 'least squares' straight line is established through the plots. This 'least squares' line relates ground snow load to elevation. Figure 3.3 is an example of the two case study plots for Mammoth Lakes, CA. The upper plot is for the nearest six stations and the lower plot is for all stations within 25 miles (40.23 km). The two plots suggest that the ground snow load should be about 300 lb/ft^2 (14.37 kN/m^2) at an elevation of 8090 ft (2466 m).

Site-specific case studies are code mandated where values cannot be obtained from the ASCE 7-05 Figure 7-1 map (Fig. 3.1). Until recently, CRREL provided this service free of charge and on request. CRREL no longer provides this valuable service. Base data on ground snow loads and snow depths for case study analysis can be obtained from the National Weather Service (www.ncdc.noaa.gov/oa/ncdc.html) and other sources, such as state water resource agencies that measure the snow water content.

The Structural Engineering Associations (SEAs) for various states may have recommendations of engineers who are capable of providing the required ground snow load case study analysis. In some states, the SEAs have published localized data for ground snow loads for their state and/or regional areas within their state. These data are not 'pinpoint' site specific, but are more 'locale' specific. It is intended to satisfy the local jurisdictional authority's case study code requirements. Tobiasson et al.[7] have done a case study for every town in New Hampshire and developed a way to modify that value for other elevations within each town. Sack and Sheikh-Taheri[8] provided a snow load map for Idaho.

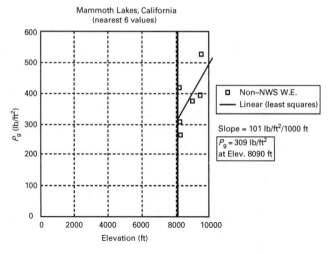

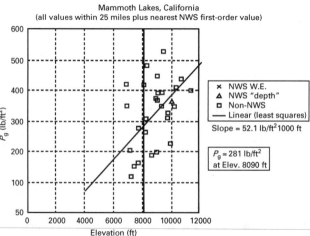

3.3 Case study (CS) plots for Mammoth Lakes, CA. Ground snow load estimation was 300 lb/sf (14.37 kN/m²) at 8090 ft (2465.83 m) elevation (Matrix IMA Archive: CRREL document).

3.2.5 Weather data

The US National Weather Service (NWS) has 204 first-order stations that record both snow depth and ground snow load. Only ground snow depth is recorded at about 11 000 other NWS 'co-op' stations, some of which are 'SNOTEL' stations that measure the water depth equivalent of the ground snow depth. In addition there are about 3300 other places across the United States where ground snow loads are measured a few times each winter by other agencies and companies. The Western Regional Climate Center (www. wrcc.dri.edu/CLIMATEDATA.html) has historical weather data for the western United States where extreme snow depth at many stations has been recorded for the past 50 to 60 years.

As an example, a project in mountainous Idaho is to be designed. By IBC/ASCE 7-05 a case study is required. The local building department ordinance requirement is a 125 lb/ft^2 (5.99 kN/m^2) flat roof snow load, P_f. As an ad hoc check on the local load requirement, you find a 'co-op' site 3 miles away from your project and 500 ft (152.4 m) higher. The extreme snow depth recorded was 10 ft (3.048 m). Based on discussions with local residents and a review of topography features and storm tracks you estimate that the extreme ground snow depth at your project will be approximately 8 ft (2.44 m).

In order to calculate ground snow load, snow density must be determined. Sack and Sheikh-Taheri [8] determined that, in Idaho, the ground snow load (P_g) for depths greater than 22 in (0.56 m) equation is

$$P_g = 2.36h - 31.9 \quad (P_g = \text{lb/ft}^2, h = \text{inches of snow depth})$$

Therefore

$$P_g = 2.36(96) - 31.9 = 194.7\,\text{lb/ft}^2(9.32\,\text{kN/m}^2)$$

Next, convert the P_g load to nominal flat roof design load (P_f) and compare it to the 125 lb/ft^2 (5.99 kN/m^2) building department ordinance requirement. Nominally $P_f = 0.7\ P_g$, assuming factors for exposure, thermal, and importance are all equal to 1.0 per ASCE 7-05. The P_f calculation is thus

$$P_r = 0.7(194.7\,\text{lb/ft}^2) = 136.3\,\text{lb/ft}^2(6.52\,\text{kN/m}^2)$$

In comparison to the building department P_f load of 125 lb/ft^2 (5.99 kN/m^2), the ad hoc load is 9% greater, and reasonably corresponds to the building department code minimum as the project snow depth estimation was not precise.

The ad hoc findings should be discussed with the local building official. Discuss the derivation of the 125 lb/ft^2 (5.99 kN/m^2) code flat roof snow load. If this load requirement is just an arbitrary value, local custom, etc., then a ground snow load case study should be obtained. On the other hand, if the load is based on 'locale' case study analysis by a reputable source (as per the discussion in Section 3.2.4), then the ad hoc analysis has basically confirmed that the code flat roof snow load is an appropriate minimum. Furthermore, the building official may not require a separate case study analysis for the project.

If you find that the building department's minimum flat roof snow load of 125 lb/ft^2 (5.99 kN/m^2) is appropriate, then consider a scenario where the multiplying factors are 1.0 or greater. First, convert the 125 lb/ft² (5.99 kN/m²) P_f load back to the ground snow load P_g (assumes multiplying factors are all equal to 1.0):

$$P_g = 125/0.7 = 178.6\,lb^2(8.55\,kN/m^2)$$

The project is a suburban fire station located on a wooded sheltered site and has a 3 to 12 sloped cold ventilated roof with some skylights over the lobby. The ASCE 7-05 Equation 7-1 is actually

$$P_f = 0.7C_eC_1IP_g$$

where

C_e = exposure factor = 1.2
      (terrain category B, sheltered – ASCE 7-05 Table 7-2)
C_t = thermal factor = 1.1
      (cold ventilated roof – ASCE 7-05 Table 7-3)
C_t = thermal factor = 1.0
      (skylight, warm roof – ASCE 7-05 Table 7-3)
I = importance factor = 1.2
      (essential facility fire station – ASCE 7-05 Table 7-4)

Utilizing the example ground snow load, the flat roof snow load P_f is

$$P_f = (0.7)(1.2)(1.1)(1.2)(178.6) = 198.0\,lb/ft^2(9.48\,kN/m^2)\quad(Roof)$$

and

$$P_f = (0.7)(1.2)(1.0)(1.2)(178.6) = 180.0\,lb/ft^2(8.62\,kN/m^2)\quad(Skylight)$$

This P_f roof load is 73 lb/ft² (43.5 kN/m²) greater than the building department minimum. The skylight load is 55 lb/ft² (2.63 kN/m²) greater. Per ASCE 7-05 Figure 7-2b (Fig. 3.4), a 3 to 12 sloped cold roof does not qualify for a slope reduction for a slippery roof, but since the skylight is a warm roof, per ASCE 7-05 Figure 7-2a (Fig. 3.4), a slippery surface slope reduction may be taken. The value is 0.85. This would reduce the 3 to 12 pitch skylight load to 153.0 lb/ft² (7.33 kN/m²) if the skylight snow can slide off and be free and clear of the skylight.

In conclusion, analyzing weather data in an ad hoc manner is not a substitute for a case study analysis, or utilizing the ASCE 7-05 Figure 7.1 map (Fig. 3.1) ground snow loads. An ad hoc weather analysis should only be used to 'ballpark' potential snow loads and initiate discussions with the building official as to the derivation of the code snow load. If the validity of the code snow load is in any way suspect, obtain a case study analysis even if the building official waives its requirement. If the code snow load is deemed

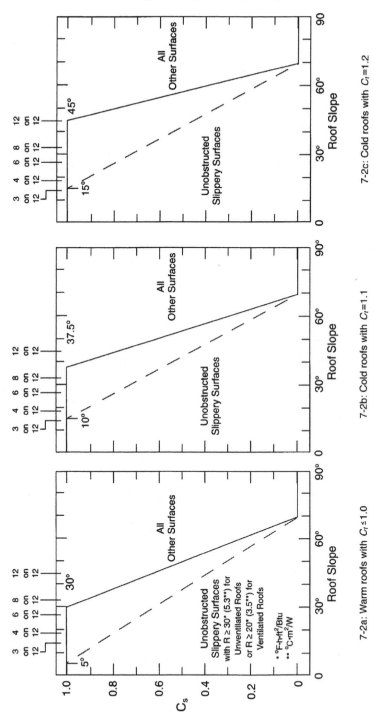

7-2a: Warm roofs with $C_t \leq 1.0$

7-2b: Cold roofs with $C_t = 1.1$

7-2c: Cold roofs with $C_t = 1.2$

3.4 ASCE Figure 7-2 – reprinted with permission from ASCE.

appropriate, be sure to utilize project-specific multiplying factors (per ASCE 7-05 Equation 7-1) to convert the minimum code snow load to the project-specific snow load.

3.2.6 General snow loading in other countries

Europe: 'Eurocode 1 –Actions on Structures –Part 1–3', CEN, 2003

There are ten climatic regions, with up to four zones in each region. Zones are used in the altitude correlation formula for each region. A 50 year MRI is used.

Japan: 'Ministry of Land, Infrastructure and Transport', 2000

These standards are very complex with multiple options for ground snow load calculations. Computer simulation modeling is encouraged. A 100 year MRI is the primary return period.

Canada: 'National Building Code of Canada', 1995, National Research Council of Canada, Ottawa

Location specific data are published. A 30 year MRI is used.

New Zealand: 'AS/NZS 1170.3', Proposed New Building Standard Department of Building and Housing, 'Snow Loads Review, Parts 1 and 2, 2007, National Institute of Water and Atmospheric Research Ltd, Christchurch, New Zealand

This report is recommending changes to current ground snow load methodology. A 150 year MRI is used.

Russia: 'SniP 2.01.07-85', 2003, "Loads and Effects", Moscow: CPPB (in Russian)*

Otstavnov and Lebedeva[9] describe the new map approach to codification of ground snow loads in Russia. A 25 year MRI is used.

Poland: 'PKNMiJ', 1980, PN-80/B-02010 (Polish standard)

Zuranski and Sobolewski[10] propose a new ground snow load standard and map utilizing a 50 year MRI.

All the counties use a ground snow load MRI annual probability. ASCE

7-05 Table C7-3 can be used to convert other MRI intervals to the 50 year MRI standard except for New Zealand.

3.3   Roof snow load per ASCE 7-05

3.3.1 Flat glazing (roof) snow loads

O'Rourke[6] gives a thorough description of the derivation of ASCE 7-05 Equation 7-1:

$$P_f = 0.7 \ C_e \ C_1 \ I \ P_g$$

as well as the C_e, C_t, and I factors and how to select appropriate values for each one per ASCE 7-05 Tables 7-2, 7-3, and 7-4. With respect to glass, Table 7-3 lists a thermal factor of 0.85 for continuously heated greenhouses with a monitored interior temperature maintained at 50 °F (10 °C) or above, and with a thermal resistance (R value) less than 2.0 °F $\times h \times$ ft^2 / Btu (0.4 K $\times$ m^2/W). This amount of heat loss then provides continuous snow melting.

The multiplying factor for a production greenhouse used for growing plants and without public access is the ASCE 7-05 Table 7-4 importance factor, I. This factor would be 0.8 for a Category I agricultural facility per ASCE 7-05 Table 1-1. If the production greenhouse is located in an exposed location (e.g. Terrain Category D) the exposure factor C_e would be 0.8. The flat roof snow load P_f would then be

$$P_f = (0.8)(0.85)(0.8) \ P_g$$
$$= 0.544 \ P_g$$

or a little over half the ground snow load.

A complete greenhouse ASCE 7-05 roof snow load evaluation procedure is given in the National Greenhouse Manufacturers Association[11] *Structural Design Manual*. This manual can be downloaded from their website: www. ngma.com.

With flat (low curb) skylights on a flat roof in snow areas over approximately 3 ft (1 m), the 0.85 thermal factor C_t should not be used. In a heated building a low curb, minimally sloped for drainage, flat skylight may have a deep snow cover over it. The heat loss through the glass will melt the bottom of the snow cover, but the internal strength of the snow will cause the snow to 'bridge' to the skylight perimeter. An airspace 'bubble' will form. The boundary between the bubble top and snow bottom can be icy, as shown in the upper diagram of Fig. 3.5. Over time, the weight of the deep snow cover above the ice boundary interface will most likely collapse the ice interface and impact the glazing. The glass/glazing design (as

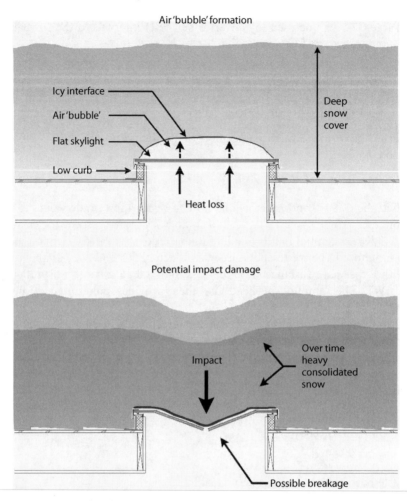

3.5 Formation of snow cover airspace 'bubble' where high heat loss occurs and its potential impact collapse on flat skylights (Matrix IMA diagram).

discussed in a subsequent chapter), besides accounting for all the ASCE 7-05 loads, should also take into account this probable impact load as well as the thermal shock effect of the snow collapsing on to the warm glass (see the lower diagram in Fig. 3.5) or its sudden exposure to subfreezing outside air. As a rule of thumb, the snow impact load is estimated by some engineers as twice the uniform load. Glass type and characteristics should be evaluated for resistance to thermal shock. Additionally, the deep and dense snow/ice on the glass/glazing will remain for a long time. The glass/glazing will be subject to static fatigue. Glass type, thickness, and other characteristics

3.6 Small pyramid skylight poking up on a flat roof of a solar mountain residence. The skylight projection is in the center portion of the roof (Matrix IMA Archive photograph).

should be evaluated and selected for potential static fatigue conditions. This is covered in a later chapter of this book.

On the other hand, a low curb pyramid or deep domed skylight on a flat roof allows a deep snow cover to be melted in a similar fashion, but the impact collapse is more like a sloughing action due to the steepness of a pyramid or deep dome sides. If a pyramid or deep domed skylight is located on a flat roof, where wind stripping is good, the tops of each may poke through the deep snow or have little snow cover, as shown in Fig. 3.6.

In another case, when designing a flat skylight on a flat roof private carport or an unheated garage, a thermal factor C_t of 1.2 would be appropriate. The importance factor I of 1.0 would also be correct for the private structure. If the carport or unheated garage were located in a 'sheltered terrain category C' location, the exposure factor $C_e = 1.1$. Consequently,

$$P_f = 0.7(1.1)(1.2)(1.0)\ P_g$$
$$= 0.924 P_g \text{ for both the roof and the skylight, or over 90\% of}$$
$$\text{the ground snow load}$$

It can be deduced that the three factors C_s, C_t, and I can greatly influence the conversion of the ground snow load, P_g, to the flat roof snow load, P_f. The appropriate coefficients must be selected with care, especially for the skylight and sloped glazing portions of the project. Frequently, the skylight/sloped glazing design snow loads will be substantially different than those for the opaque roof.

Table 3.1 Common roof slope factors

Slope	Warm: C_t 1.0	Cold: C_t 1.1	Unheated: C_t 1.2
3/12	0.85	0.95	1.00
4/12	0.80	0.90	0.95
6/12	0.65	0.80	0.75
8/12	0.55	0.60	0.65
12/12	0.45	0.45	0.45
Gutters	1.00	1.00	1.00

3.3.2 Sloped glazing (roof) snow loads

In most projects, sloped glazing occurs over a heated interior space. ASCE 7-05 permits a roof load slope reduction by Equation 7-2:

$$P_s = C_s \, P_f$$

where C_s is the slope reduction factor.

As sloped glazing is usually both an unobstructed slippery surface that will let all the snow slide off the eaves to a space that will accept all the snow and also warm roof ($C_t \leq 1.0$), the dashed line in ASCE 7-05 Figure 7-2a (Fig. 3.4), can be used. C_s from 0° to 5° slope is 1.0, and from there declines linearly to $C_s = 0$ at the 70° slope. ASCE 7-05 Figures 7-2b and 7-2c (Fig. 3.4) are for cold and unheated roofs and their slopes start at 10° and 15°, respectively. The NGMA[11] data for common sloped roof C_s factors are correlated by Table 3.1.

To return to the fire station example, in Section 3.2.5, the flat roof skylight snow load P_f is 180 lb/ft^2 (8.62 kN/m^2). The 3 to 12 sloped glazing skylight over the lobby would then have a warm roof C_s of 0.85 from Table 3.1. Additionally, it has been decided that a 6 to 12 glass canopy over the front entry walkway is necessary. Again, referring to Table 3.1, this $C_s = 0.75$ because the walkway is unheated. Therefore,

$$P_s = 0.85(180) = 153.0/, \text{lb/ft}^2 \ (7.33 \, \text{kN/m}^2) \quad \text{(Skylight)}$$

and

$$P_s = 0.75(180) = 135.0/, \text{lb/ft}^2 \ (6.47 \, \text{kN/m}^2) \quad \text{(Glazed canopy)}$$

Note that it is assumed that both the skylight and canopy shed snow into a protected landscaped area where full winter snow slide accumulation can occur and not endanger people or property. If this is not possible, the designer must consider holding the snow on the sloped glazing, as discussed later in Section 3.4.2, and not take the reduced 'unobstructed slippery surface' P_s load, i.e. the design for the ASCE 7-05 Figure 7-2 'all other surfaces' P_s snow load (see Fig. 3.4).

3.3.3 Partial snow loading on glazing

Although ASCE 7-05 allows slope reductions for glass and glazing as described in Section 3.3.2, partial and potentially very high concentrated loadings can occur. In the previous 3.3.2 example, the sloped glass skylight eave could be at the exterior wall line with no roof overhang or with a down slope roof overhang.

In the case without an overhang, the skylight snow melt water will freeze when exposed to subfreezing air at the eave edge and form icicles and an ice dam. The icicles/ice dam and ponded upslope melt water constitute a concentrated load of $2P_f$ in their own right (see ASCE 7-05, Section 7.4.5), but this eave ice buildup also prevents snow from sliding off the skylight until the static force of the retained snow breaks the ice's bond at the skylight's eave framing members. Then a roof avalanche will occur. An example of a sudden cold roof avalanche is shown in Fig. 3.7. Potential roof avalanche mitigation options are discussed in Sections 3.4.2 and 3.4.5 below.

Lepage and Williams[12] conducted cold room testing of sloped glazing with an eave snow rail to prevent snow sliding off 45° sloped glazing. They found that the snow rail had to be isolated from the building heat (kept at outdoor temperature) in order to control snow creep and icicles. If this snow retention technique is utilized on sloped glazing, then the warm roof unobstructed slippery surface slope reduction factor C_s should not be taken.

Nielsen[13], in a Norwegian study of snow loads on glass roofs, points out that on glass roofs of over 30° slope snow will slide. In his study, data were not available for slopes less than 30°. In order to promote snow sliding, Nielsen recommends glass with a high heat loss and metal glazing profiles

3.7 Sudden roof snow avalanche. Metal standing seam 4 to 12 cold roof (Matrix IMA Archive photograph).

that do not act as snow fences nor allow melt water to freeze on them, i.e. no better insulated than the glazing itself. Nielson also notes that these recommendations are contradictory to energy conservation needs. Unfortunately, a paradoxical problem results: high heat loss to encourage snow/ice melting versus low building energy usage. The designer needs to balance the project's needs.

In the example of a 3 to 12 (14°) sloped skylight, the snow will slide, but generally slowly, i.e. creep down the skylight. If the snow is deep and has some compaction and reasonable internal strength, the snow slab creep can cantilever beyond the eave a substantial distance before calving (breaking off), like a glacier does. This snow calving poses a serious risk to whatever may be below. Refer to Fig. 3.8 for a photo of a substantial snow cantilever. To avoid this situation the snow must be held on the roof as discussed in Section 3.4.2 below.

In the case of a rough surface roof, or a nonslippery roof overhang downslope from a low height skylight eave curb, the icicles/ice dam condition can occur on this low height skylight eave curb. Sliding skylight snow will probably pile up at the skylight eave curb as the roof snow below the eave curb sill will likely stop the slide. Skylight snow melt water will either freeze at the skylight eave curb (forming ice/ice dams) or flow under the eave piled-up snow, and then under the roof snow pack to the roof eave where again ice dam/icicles will form.

3.8 Snow creep cantilever. Metal standing seam 2 to 12 cold roof. Note: ripped away snow fence bars (Matrix IMA Archive photograph).

3.9 Fully sheltered sloped glazing entry below main roof rake (Matrix IMA Archive photograph).

For ice dam/icicle conditions, skylight eave glazing should be designed for increased loading of $2P_f$ (per ASCE 7-05 Section 7.4.5) as well as the full and half balanced partial loading scenarios per ASCE 7 - 05 Section 7.5. This partial loading analysis would be especially important if a single glass panel within the skylight was sheltered at the upslope end (not subject to uniform snow load) and fully loaded at the downslope portion. O'Rourke[6] discusses with examples the ASCE 7-05 methodology of partial loading for continuous structural members such as roof purlins in metal building systems. Continuous skylight rafters and/or mullions may require the same type of engineering analysis. The glass panel design for partial loading is discussed in a subsequent chapter of this book. Figure 3.9 shows a fully sheltered sloped glazing/skylight. This skylight is only exposed to wind-driven snow.

3.3.4 Drifts on to lower level glazing

O'Rourke[6] discusses with examples the ASCE 7-05 methodology for determining wind-drifted unbalanced snow loading on sloping roofs and wind drift surcharge loads at stepped (lower) roofs. An example of extreme drifting/unbalanced load on a gable roof is shown in Fig. 3.10.

3.10 Massive unbalanced drift snow on north facing 5 to 12 warm roof. Note: sun and wind from south (Matrix IMA Archive photograph).

Obviously, skylight or sloped glazing located in an area subject to wind-drifted unbalanced loading or within a drift surcharge zone of a lower stepped roof would need to be designed for this increased loading. Be sure to check the weather data for both storm winds and nonstorm wind directions. Many times the directions can be diametrically opposed to each other and drift locations will change radically. Depending on the project's size and complexity, computer analytical modeling or wind tunnel/water flume modeling of snow drifting should be considered. Many published papers by respected researchers exist on the use of these techniques. Some examples are: Irwin *et al.*[14] for analytical modeling, Isyumov and Mikitiuk[15] and Dufresne de Virel *et al.*[16] for wind tunnel modeling, and O'Rourke and Wrenn[17] for water flume modeling.

3.3.5 Roof projections

O'Rourke[6] discusses with examples the ASCE 7-05 drifting snow loads associated with parapets and rooftop projections. Again, a skylight or sloped glazing located near or adjacent to any kind of roof projection would need to be designed for this surcharge drift loading.

3.3.6 Sliding snow

ASCE 7-05 requirements state that snow sliding off a sloped roof on to a lower roof shall be determined for slippery upper roofs (slope $> \frac{1}{4}$ to 12, or $\sim 2\%$). The sliding load is distributed evenly over a lower roof for a width of 15 ft (4.57 m), or proportionally reduced when the lower roof is less than 15 ft (4.57 m) wide. Furthermore, the height between the eave of the upper roof measured directly down to the lower roof is taken into consideration. The sliding snow surcharge depth cannot exceed this dimension. ASCE 7-05 Equation 7-3 for snow density is used to calculate the height of the sliding snow surcharge plus the uniform load on the lower roof.

The ASCE 7-05 Equation 7-3 is

$$Y = 0.13\,P_\mathrm{g} + 14 \leq 30\,\mathrm{lb/ft}^3 \quad (Y = \mathrm{lb/ft}^3,\ P_\mathrm{g} = \mathrm{lb/ft}^2)$$

The density Y maximum is 30 lb/ft^3 (480 kg/m^3). This maximum density value was quantified during roof drifting research by Speck[18]. This equation is only valid for calculating roof sliding or drift snow density.

If the sum of the sliding snow surcharge height plus the uniform snow height exceeds the actual height, sliding snow blockage occurs and the sliding snow surcharge load is reduced to conform to the actual height. Examples by O'Rourke[6] clearly show the application of the ASCE 7-05 Section 7.9 equation for sliding snow:

$$S_\mathrm{L} = 0.4 P_\mathrm{f}\,W$$

where W = horizontal distance eave to the ridge in ft.

Although the ASCE 7-05 Commentary, Section C7.9, mentions the dynamic effect sliding snow may have on the lower roof, no mention is made of deep, dense, upper roof snow glacially creeping off the eave and cantilevering. These chunks of snow then calve off and impact the roof below – often quite dramatically. Estimating the dynamic impact force would be problematic at best, due to the multiplicity of variables involved with the calving snow/ice chunks. A 'rule of thumb' estimation of the dynamic load used by some engineers is twice the static snow load. A more precise method would be to use computer modeling to determine dynamic design load criteria. The Fig. 3.11 photo is a condition where a small greenhouse glazed roof failed under the impact. The cantilevered snow fell only about 24 inches (0.6 m). The upper roof was a slippery standing seam metal cold roof with snow fences that failed under the glacial sliding. This occurred during the heavy snow/rain event of spring 2006 in the western United States.

The best defense to protect a skylight or sloped glazing from sliding snow is to place the glazing above or away from this danger. Where this is not possible, and the upslope roof eave is above the sloped glazing/skylight, snow arrestors or fences should be employed, but they must have adequate structural design to retain the snow on the upper roof. An example is shown in Fig. 3.12. If this is done, the upper sloped roof should be designed for the ASCE 7-05 Figure 7-2 (Fig. 3.4) 'all other surfaces' sloped roof snow load P_s.

3.11 Snow cantilever calving impact damage to sloped glazing/ greenhouse 2 to 12 roof. Note: snow inside enclosure (Matrix IMA Archive photograph).

3.12 Structurally designed snow arrester. Note: heat tracing in gutter not powered at time of photo (Matrix IMA Archive photograph).

3.4   Other roof snow glazing issues

3.4.1 Snow bridging

As diagrammed in Fig. 3.5, snow can bridge over a skylight due to the heat loss through the glazing. The bridging snow load is transferred internally within the snow to the colder perimeter of the skylight. Although the diagram indicates a flat skylight, this bridging can also occur on a moderately sloped skylight.

3.4.2 Snow/ice retention

In some design circumstances, snow/ice retention systems must be used to prevent snow/ice from sliding off a roof. This is especially important if the sliding snow/ice will fall into an area where harm to people or damage to property is likely. Sliding snow/ice falling on to a skylight/sloped glazing, as depicted in Fig. 3.11, is one such instance where no one was hurt, but a significant amount of property damage did occur.

The upper metal roof in Fig. 3.11 had snow fences installed that attached to the standing seams. This was a woefully inadequate structural solution, as these fences were torn away by the roof snow glacial creep, as depicted in Fig. 3.8. As Tobiasson et al.[19] explain, metal roofs need to expand and contract. These roofs effectively 'float' over the underlying substrate. Most snow fences, clips, or guards are secured to the top surface of the metal roof. This allows roof thermal movement. If these snow retainage devices are fastened through to the underlying roof structure, the fasteners will inhibit the thermal movement of the metal roof. Furthermore, fastener penetration holes provide a path for water leakage into the underlying materials. A common plastic snow guard is depicted in Fig. 3.13. Tobiasson et al.[19] indicate that 'improved design guidelines, standards and performance criteria are needed for snow guards on metal roofs'. Today, over a decade later, that statement still rings true.

A structurally designed snow arrester as shown in Fig. 3.12 would have prevented the glacial snow/ice movement shown in Fig. 3.8. As a structurally designed snow arrester needs to resist the downslope component of the vertical snow load, the arrester may need to be anchored into the primary structural roof support members. The design challenge in doing this with a high thermal movement roof covering is to detail a waterproof flashing system where the snow arrester penetrates the roof covering and still allows for high thermal movement to occur. Figure 3.14 is an example detail of a structural snow arrester that addresses the thermal movement concerns. Mackinlay and Flood[20] designed the retrofit eave structural snow arrester shown in Fig. 3.12.

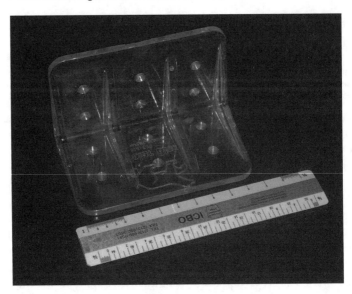

3.13 Typical plastic snow guard. Note: the countersunk fastener locations designed for screw attachment where adhesive attachment is not used (Matrix IMA Archive photograph).

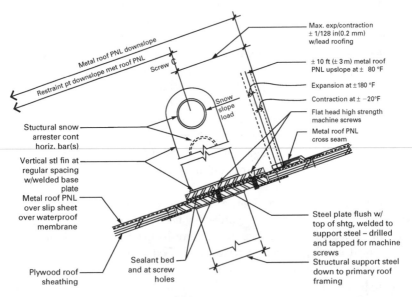

3.14 Example detail of structural snow arrester on standing seam metal roof designed to accommodate metal roof thermal movement (Matrix IMA diagram).

3.4.3 Ice dams/icicles

Since sloped glazing/skylights are warm roofs, snow will usually slide if the slope is over 30°. Where the slope is shallower, snow can accumulate on the glass. Melt water will be generated at the snow/glass interface. This melt water will run down the slope and refreeze at the skylight eave when the melt water is exposed to subfreezing outside air and/or cold eave framing members. Ice dams/icicles will form. These ice dams/icicles can act as snow arresters and cause more snow to accumulate on the upslope glazing.

At some point, the static force of the upslope retained snow will break the ice bond to the glazing caps and/or other eave surfaces. This usually happens suddenly, and the ice breaking loose can cause extensive damage to these building components. Concurrently, the retained roof snow will probably avalanche off the roof, taking with it the ice chunks from the ice dam/icicles. Large ice dams as shown in Fig. 3.15 can occur even on slippery surface roofs.

3.15 Large ice dam at cold roof overhang. Note: primary icicles have been removed (Matrix IMA Archive photograph).

3.4.4   Sliding snow/avalanche

A sliding snow/roof avalanche can cause severe damage to roof projections such as plumbing and gas vents, chimneys, and lightning rods. Light tubes, small skylights, and small roof windows set on high chimney-like curbs are also at high risk. These types of projections can easily be sheared off or bent or otherwise severely damaged by either a glacial creep snow movement or a sudden roof snow avalanche. Devices such as roof crickets or snow splitters should be used to cleave and divert the snow slide around the sides of the projection.

With a sloped glazing/skylight condition, the design probably will not have plumbing or gas vents penetrating the glazing. On the other hand, a custom fireplace metal chimney could dramatically penetrate a glass roof. If so, the sliding snow load on the chimney should be ascertained. Potentially, the designer could entertain thoughts of using the glass edge at the chimney penetration hole to resist the sliding load on the chimney. This should be avoided as glass edge loading causes glass stress fracturing (see Fig. 3.16, upper diagram). The chimney should be independently braced, as shown in the lower diagram of Fig. 3.16.

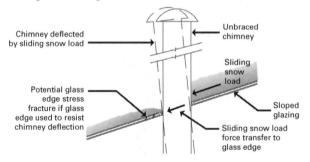

Glass edge loading damage

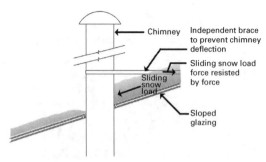

Independently braced chimney

3.16 Potential glass breakage if glass edge used to resist sliding snow load on chimney. The chimney should be independently braced (Matrix IMA diagram).

If lightning rods are required on the sloped glazing/skylight framing members at various locations, a rigid rod will have a high likelihood of being sheared off. A spring loaded/flexible/pivoting rod that can lay flat under the moving snow has a higher survivability factor during sliding snow/ avalanche events.

3.4.5 Melt water drainage

As indicated previously, melt water running under a snow pack, once it reaches subfreezing air or cold surfaces, will refreeze and form ice dams/ icicles. In order to mitigate the ice dam/icicle formation, the melt water must be kept liquid as long as possible when encountering these freezing conditions, or it needs to be drained away prior to its contact with these freezing roof elements.

Snow packs on nominally flat skylights on a flat roof will drain melt water over their low curbs on to the surrounding roof that is snow covered. The melt water then flows under the insulating snow pack to roof drains that are piped down through the heated interior space. The piping is insulated to prevent condensation on the pipe's exterior surface.

With sloped glazing/skylights, the melt water should be drained off into a heat-traced gutter and downspout system prior to the melt water's exiting from under the insulating snow pack. If the glazing is steep and the glass warm, the snow pack will slide off relatively frequently provided the glazing caps and gutter lip do not impede the slide. With very cold temperatures, any moisture on the sloped glass will freeze and form ice sheets on the glass. After a small amount of time the warm glass will create a thin melt water layer below the ice sheet. When that happens, the ice sheets will slide, and generally, if thin, they will break up and flutter in the air until they land. This situation still can be dangerous due to the sharp ice sheet edges. Sliding thin ice sheets can be retained at the glazing eaves by the use of a heat-traced perforated metal lip extending approximately 1 inch (25.4 mm) above the glass. The warm perforations allow the melt water to drain into a heat-traced gutter.

If a snow fence or arrester is placed just downslope of the melt water heat-traced gutter, the snow pack on the sloped glazing/skylight will be retained. The snow pack covers the gutter and helps insulate the gutter melt water from the subfreezing outside air. As Lepage and Williams[12] demonstrated, the snow fence/arrester must be kept below freezing to prevent the retained snow pack from extruding around the snow fence/arrester. The ice over the gutter in Fig. 3.12 occurred because the gutter heat-tracing was not yet functional at the time the photo was taken. Buska et al.[21] summarize the difficulties associated with the use of electric heating systems for combating icing problems on metal roofs.

In the current marketplace there are some electrically heated eave flashing systems. These systems can substantially mitigate eave icicle formation, but ice dam formations are likely to occur just upslope of the heated eave. If this 'just upslope' zone is the warm sloped glazing/skylight eave framing with warm glass, the potential ice dam formation may well be substantially mitigated, except when a power outage occurs.

In summary, the key to mitigating ice dams/icicle formation is to keep the melt water liquid throughout its entire drainage path to underground drainage collection systems.

3.5   Vertical glazing snow issues

3.5.1   Ground snow depth

Windows located near grade are susceptible to lateral pressures from the depth of the ground snow. Drifting snow can cause a significant snow depth increase against the side of a building over and above a uniform ground snow depth. O'Rourke[6] states that ASCE 7-05 remains silent on the issue of snow lateral pressure. He also indicates that a somewhat conservative but general approximation of the lateral load can be calculated by utilizing Rankine's theory for cohesionless soil and a frictionless wall. O'Rourke[6] provides an example of this calculation.

Rice and Dutton[22] indicate that snow load lateral force was considered in the design of the glass façade for the La Cite de Sciences et de La Villette,

3.17 Snow up against building. Note: top of window just above ground snow. Window head is ± 10 feet (3.048 m) above grade (Matrix IMA Archive photograph).

3.18 Snow/ice curl at eaves of sloped metal standing seam warm roof (Matrix IMA Archive photograph).

Paris, France. Their lateral force rationale was not disclosed. A window in deep snow that is subject to a lateral snow force is shown in Fig. 3.17.

3.5.2 Roof snow/ice curl

Glacial creep of snow off a roof eave overhang in many instances curls down and back under the overhangs shown in Figs 3.18 and 3.20. This snow curl has ice/icicles at its leading edge. With short roof overhangs this snow/ice curl can press against windows and break the glass. As shown in Fig. 3.19, overhangs should be extended far enough so that this snow/ice curl is not likely to pose a threat to the windows. The windows in Fig. 3.19 are high up on the wall and fully sheltered from snow/ice curl. They will not be damaged.

3.5.3 Roof snow/ice slide or avalanche

Sliding roof snow lands on the ground snow generally outboard of the roof eave. Over time a snow berm will form. This snow berm is dense and compacted. It can be further hardened when roof melt water and dripping icicle water freezes on and in the snow berm.

When ice dams/icicles are broken loose from the roof eave, this hard ice is likely to fall and bounce off the sloped side of the berm and impact the exterior wall. A soccer ball size chunk of ice can punch a hole in siding and break studs. It will impact a window with such force that the glass will shatter and cause other interior damage as it flies and bounces across the room (see the schematic in Fig. 3.20). In this situation, use of heavy mesh

3.19 Snow/ice curl effects mitigated with sufficient cold roof overhang and bracing (Matrix IMA Archive photograph).

detachable screens to protect the windows during the winter is a design option that should be considered.

3.6   Conclusions

Ground snow load is the basis of all snow load roof and sloped glazing/ skylight load design. Verifying that the source data are appropriate and adequate is of prime concern. Independently cross check the snow load source data for case study extreme snow load locations, especially where the data source is the local jurisdictional authority.

Calculate the roof snow loads per ASCE 7-05. It is recommended that O'Rourke[6] should be used as a guide to making the load calculations for all roof and sloped glazing snow loading conditions. Utilize the NGMA[11] *Structural Design Manual* as appropriate for sloped glazing and green-houses. Additionally, do not reduce the thermal factor (C_t) to less than 1.0

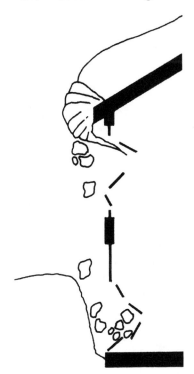

3.20 Diagram of upper window breakage due to ice curl. Lower window breakage due to falling ice chunks bouncing off snow/ice berm (Matrix IMA Archive diagram).

unless the project can specifically meet all the greenhouse criteria contained in ASCE 7-05 Table 7-3. Do not reduce the importance factor (I) to less than 1.0 unless the project qualifies as a low hazard to human life occupancy, per ASCE 7-05 Table 7-4, which then references ASCE 7-05 Table 1-1.

Snow on sloped glazing will creep or slide on slopes as little as 2%. On slopes of 30% or greater, snow should slide off the sloped glass readily, unless retained in some manner. Between a slope of 2% and 30% the snow will creep at low slopes and increase its movement at steeper slopes, again unless retained in some manner. When the snow pack over sloped glazing moves slowly or is retained, the glazing material (primarily glass) is potentially subject to long-term static fatigue and partial-area loading conditions. These conditions should be considered in the glazing selection process, as discussed in a later chapter.

Extreme snow loading can occur in areas with low, 50-year mean recurrence interval (MRI) ground snow loads. A single storm event could easily increase the design roof or sloped glazing/skylight snow load by a substantial percentage. For example, in the January 2007 ice storm,

Ardmore, OK, experienced one inch (25.4 mm) of ice. At 4.7 lb/ft^2 (0.23 kN/m^2), this weight approaches the 50 year MRI of 5 lb/ft^2 (0.24 kN/m^2) and is approximately 134% of the normal flat roof design snow load. Although the load increase is low, the percentage increase is large. On the other hand, the January 2008 three day storm dumped 11 ft (3.35 m) at the Kirkwood, CA, ski area. This wet snow, commonly called 'Sierra cement', has a weight of approximately 12.5 lb/ft^3 (200 kg/m^3). This single storm event added an approximately 137 lb/ft^2 (6.56 kN/m^2) load. This single storm load added about 46% of the local flat roof design load requirements.

Concurrent with the roof and sloped glazing/skylight loading analysis, the physical aspects of the snow must be dealt with. Is snow retained on the roof or skylights? If it slides off, where does it fall? Will ice dams/icicles form? Can they be mitigated? How will the melt water be drained? Will curling/falling snow and ice cause damage to roofs, eaves, walls, or windows? Will such events harm people or other property? All such questions must be resolved during the project's design process. The design snow loading analysis should then reflect the decisions made.

3.7   References

1. International Code Council Inc., *2006 International Building Code*, ICC Inc., Country Club Hills, IL, 2006.
2. International Code Council Inc., *2006 International Residential Code*, ICC Inc., Country Club Hills, IL, 2006.
3. American Society of Civil Engineers, *Minimum Design Loads for Building and Other Structures*, ASCE 7 - 05, ASCE, Reston, VA, 2006.
4. O'Rourke, M., Koch, P. and Redfield, R., *Analysis of Roof Snow Load Case Studies: Uniform Loads*, CR83-01, CRREL, Hanover, NH, 1983.
5. Tobiasson, W. and Greatorex, A., Database and methodology for conducting site specific snow load case studies for the United States, in Third International Conference on *Snow Engineering*, Balkema, Rotterdam, NL, 1997.
6. O'Rourke, M., *Snow Loads: Guide to the Snow Load Provisions of ASCE 7-05*, ASCE, Reston, VA, 2007.
7. Tobiasson, W., Buska, J., Greatorex, A., Tirey, J., Fisher, J. and Johnson, S., *Ground Snow Loads for New Hampshire*, TR-02-6, ERDC/CRREL, Hanover, NH, 2002.
8. Sack, R. and Sheikh-Taheri, A., *Ground and Roof Snow Loads for Idaho*, Department of Civil Engineering, University of Idaho, Moscow, ID, 1986.
9. Otstavnov, V. and Lebedeva, I. The new map of ground snow loads for the Russian building code, in Fifth International Conference on *Snow Engineering*, Balkema, Leiden, NL, 2004.
10. Zuranski, J. and Sobolewshi, A. Ground snow loads in Poland, in Fifth International Conference on *Snow Engineering*, Balkema, Leiden, NL, 2004.
11. National Greenhouse Manufacturers Association, *Structural Design Manual*, NGMA, Harrisburg, PA, 2004.
12. Lepage, M. and Williams, C., 'Cold room studies for snow and ice control on

buildings, in First International Conference on *Snow Engineering,* SR89-6, CRREL , Hanover, NH, 1989.

13. Nielson, A., Snow melting and snow loads on glass roofs, in First International Conference on *Snow Engineering,* SR89-6, CRREL, Hanover, NH, 1989.
14. Irwin, P., Williams, C., Gambel, S. and Retziaff, R., Snow load prediction in the Andes Mountains and downtown Toronto – FAE simulation capabilities, in Second International Conference on *Snow Engineering*, SF92-27, CRREL, Hanover, NH, 1992.
15. Isyumov, N. and Mikitiuk, M., Wind tunnel model studies of roof snow loads resulting from muliple snowstorms, in Third International Conference on *Snow Engineering*, Balkema, Rotterdam, NL, 1997.
16. Dufresne de Virel, M., Delpech, P. and Sacre, C., Wind tunnel investigations of snow loads on buildings, in Fourth International Conference on *Snow Engineering*, Balkema, Rotterdam, NL, 2000.
17. O'Rourke, M. and Wrenn, P., Water flume evaluation of snowdrift loads on two-level flat roofs, in Third International Conference on *Snow Engineering*, Balkema, Rotterdam, NL, 1997.
18. Speck Jr, R., *Analysis of Snow Loads Due to Drifting on Multilevel Roofs*, MS Thesis, Department of Civil Engineering, Rensselaer Polytechnic Institute, Troy, NY, 1984.
19. Tobiasson, W., Buska, J. and Greatorex, A., Snow guards for metal roofs, in 8th Conference on *Cold Regions Engineering,* ASCE, Reston, VA, 1996.
20. Mackinlay, I. and Flood, R., The impact of ice dams on buildings in snow country, in Third International Conference on *Snow Engineering,* Balkema, Rotterdam, NL, 1997.
21. Buska, J., Tobiasson, W., Fyall, W. and Greatorex, A., Electric heating systems for combatting icing problems on metal roofs, in Fourth International Symposium on *Roofing Technology,* NRCA, Rosemont, IL, 1997.
22. Rice, P. and Dutton, H., *Structural Glass,* Taylor & Francis, London, UK, 1995.

4

Architectural glass to resist snow loads

R. H. DAVIES and N. VIGENER, Simpson Gumpertz &
Heger Inc., USA

Abstract: Glass design to resist snow loads includes safety and serviceability
review of sloped, vertical, and overhead glazing systems where water, snow,
and ice accumulation are possible. Sloped glazing systems must resist long
duration loading, impact damage, and water penetration, while also
avoiding ponding water and sliding snow or ice hazards. Glass and glazing
industry organizations have established guidelines and standards for the
design of sloped glazing. Building codes require load combinations
including snow and ice loading on sloped glazing systems for structural
design. The American Society of Testing and Materials (ASTM) standard
E1300, *Standard Practice for Determining Load Resistance of Glass in
Buildings*, describes procedures to determine load resistance and deflection
for monolithic, laminated, and sealed insulating glass (IG) unit combinations under uniform, short, or long duration lateral loads. E1300 annexes
provide mandatory information and appendices provide voluntary
procedures and background information. Problem examples in the chapter
describe standard practice procedures for various assembly combinations
and load durations, including solutions for strength and serviceability.
Discussion extends consideration of glass strength and serviceability,
proposed load duration effects on heat-treated glasses, and time and
temperature effects on laminated assemblies. Problem examples in the
chapter describe procedures beyond standard practice for various assembly
combinations to further develop an understanding of strength and
serviceability.

Key words: sloped glazing, standard practice, load resistance, load duration
factors (LDF), probability of breakage, static fatigue, interlayer stiffness,
temperature effects.

4.1   Introduction

This chapter describes glass design to resist snow loads, including safety and
serviceability strategies for sloped, vertical, and overhead glazing assemblies. This chapter includes the following:

- Sloped glazing design strategy for snow and ice accumulation and to avoid the risk of water penetration
- Description of United States Standard Practice to determine load resistance
- Examples of United States standard practice procedures to determine acceptable performance
- Discussion on static fatigue effects on glass strength, load duration, and temperature effects on laminated glass assemblies
- Examples of nonstandard design approaches, providing selected reference resources for additional information

4.2   Sloped glazing system design strategy for snow and ice

Functionally, sloped glazing systems are roofs. They are exposed to more intense precipitation than vertical glazing and require competent water-proofing design. In addition, sloped glazing systems subject to snow loads are vulnerable to the effects of long duration loading, impact damage, and water penetration. Vertical and overhead assemblies may be subject to similar conditions. Design approaches to address structural and service-ability demands posed by snow accumulation on sloped glazing systems are discussed later in this chapter. These demands require careful design consideration to achieve acceptable performance. Competent glazing systems must have adequate glass and framing structural resistance, system slope, and drainage design under all conditions in all seasons.

Sloped glazing assemblies are more vulnerable to impact damage and water penetration than typical roofing assemblies. Traditional designs utilize relatively steep slope to ensure rapid drainage and to limit snow accumulation. Considerations of glazing geometry, detailing, and relative placement within a building remain the most obvious and effective strategies to cope with the effect of snow accumulation on sloped glazing, namely by preventing it from occurring in the first place. They include the following:

- Avoid locating glazing assemblies in areas that are subject to sliding snow and ice (Fig. 4.1). These include locations below the eaves of adjacent higher roofs or other skylights, or along the base of building elevations that are subject to accumulation.
- Because glazing has a low insulating value compared to roof assemblies with dedicated insulation and because exterior glazing surfaces are relatively slippery, significant amounts of snow and ice do not accumulate on sloped glazing systems with sufficient slope. Structural load calculations performed in accordance with model building codes

4.1 Avoid locating glazing assemblies in areas subject to sliding snow and ice.

acknowledge this effect. A slope of 3 in 12, or approximately 15°, is a prudent minimum for sloped glazing.

- The same properties (relatively low insulating value, slippery surfaces, steep slope) may cause ice and snow that accumulates under relatively cold conditions (e.g. during the night) to slide off the glazing when temperatures increase, which can cause damage to other building components or injury to pedestrians. Applied heat, such as an electrical heat trace system, that prevents ice accumulation provides an approach to prevent this problem.

- Sloped glazing systems are typically part of a larger roof assembly that includes other steep or low-slope roofs, gutters, and rising walls. These features must be designed to withstand not only the expected structural loads, but also the waterproofing demand posed by the significant volume of sliding snow accumulation below the base of the sloped glazing. For example, the waterproofing height of gutters or drainage troughs must be sized based on an anticipated height of snow accumulation, which must be determined from the geometry of the roof and glazing, volume of expected snow fall, and density of the snow (Fig. 4.2). This is an important design consideration because rainfall on a snow blanket may cause water to pond against the glazing unless the waterproofing height below the glazing is sufficiently high (Fig. 4.3).

Effective waterproofing design for sloped glazing is an indispensable component of designing for snow loads because accumulated snow and ice will eventually melt and the water will drain off the glazing. United States model building codes do not contain specific descriptive design requirements to ensure adequate waterproofing performance of sloped glazing assemblies. Skylight designers and specifiers must refer to standards

4.2 Gutters and drainage trough waterproofing height must be sized based on an anticipated height of snow accumulation.

4.3 Rainfall on a snow blanket may cause water to pond against glazing unless waterproofing height is sufficiently high.

and guidelines established by industry associations and standard-writing organizations for minimum performance requirements.

4.2.1 AAMA resources

The American Architectural Manufacturers Association (AAMA) developed a system of performance classifications for fenestration systems that is based on performance under ASTM (see below) standard tests. These classifications aid the designer in selecting fenestration systems based on structural demand, air and water penetration resistance, and subjective judgment of anticipated service and use (e.g. 'residential' or 'commercial'). At this time, AAMA's classification system does not include sloped glazing. AAMA does, however, publish a document, *Glass Design for Sloped Glazing* (AAMA, 1987), that provides a useful background discussion on thermal stresses, deflection of framing members, and drainage of water.

4.2.2 ASTM testing methods

The American Society of Testing and Materials (ASTM) provides testing methods for assessing the air infiltration and water penetration of building enclosure systems (ASTM, various dates), including criteria used to assess acceptable performance. This chapter discusses the ASTM standard for determining glass resistance to loads only.

4.2.3 The Whole Building Design Guide

The Whole Building Design Guide (WBDG) (Vigener and Brown, 2006), prepared by the National Institute of Building Science (NIBS) in Washington, DC, contains practical design guidance to improve the waterproofing performance of sloped glazing systems. Similar to prudent vertical glazing design, sloped glazing design must both limit the amount of water penetration into the system and provide a system to drain the inevitable water penetration back to the exterior without causing leakage to the interior. The recommendations in the WBDG include, in part, the following, which are paraphrased from the material on the WBDG website:

- Provide a continuous system of gutters, integral with the rafters and cross members, to collect leakage and condensation. Water collected within the system must be drained from gutter to gutter, and eventually to the sill flashing and to the exterior. Continuous skylight rafters, which have continuous gutters, provide more reliable waterproofing performance than discontinuous individual sections that are spliced together because the splices rely on sealant or membrane patches, which tend to deteriorate over time.

- Provide continuous sill flashing. As for vertical glazing, continuous and durable sill flashing that is sloped to drain to the exterior is one of the most important aspects of waterproofing design.
- Provide an exterior wet seal, rather than dry gaskets. A wet seal consisting of noncuring butyl glazing tape and an exterior silicone cap bead, installed with proper workmanship, provides better waterproofing performance than a dry gasket.
- Provide unobstructed drainage over the glazing. Projecting horizontal pressure bars that buck water are less reliable than flush-glazed horizontal mullions.

4.3   Structural codes and standards

For snow and ice loading, United States local jurisdictions increasingly reference the provisions of the *International Building Code* (ICC, 2006) and/ or American Society of Civil Engineers standard, *Minimum Design Loads for Buildings and Other Structures* (ASCE, 2005).

4.3.1 Snow and ice load determination

Codes determine snow loading as a function of geographical location and associated ground snow load (based on measured historical weather data representative of the general geographical location), roof (or glazing) configuration and slope, R value (a measure of insulating value) of the roof or glazing assembly, exposure category (a measure of the landscape characteristics near the building), building importance, and other factors. The standard determines ice load as a function of geographical location and associated ice thickness, component surface area, exposure category, building importance, and other factors. Based on requirements, designers must review snow loads in combination with other loads listed below, including special considerations to avoid or design for falling or sliding snow impacts. The following loads must be taken into account:

- Dead loads (including self-weight loading)
- Live loads (for accessible floor or roof areas)
- Roof live loads (including maintenance loading)
- Rain loads (including ponding loads if unavoidable)
- Wind loads (typically as components and cladding loading)
- Temperature loads (including extreme low and high temperatures)
- Seismic loads (typically as components and cladding loading)

Designers must review applicable local codes to determine prescribed specifications for glass products and assemblies. Most codes require designs that prevent broken overhead glass from falling from the glazed opening, specifying the use of laminated glass, wired glass, or mesh screens at the building interior side. Designers must provide systems that protect occupants from or eliminate the potential for falling glass.

4.3.2 Material and product standards

Many codes refer to American Society of Testing and Materials (ASTM) standards for glass materials and standard practices. In addition, industry organizations such as the American Architectural Manufacturer's Association (AAMA), Glass Association of North America (GANA), and Insulating Glass Certification Council (IGCC) provide information on products, quality standards, and guidelines.

4.4   Glass specification per United States standard practice

Glass kind, thickness, and assembly specification depends on the glazing system configuration, code requirements, and other factors. ASTM E1300, *Standard Practice for Determining Load Resistance of Glass in Buildings* (ASTM, 2007), describes procedures 'to determine load resistance of specified glass types, including combinations of glass types used in a sealed insulating unit, exposed to a uniform lateral load of short or long duration, for a specified probability of breakage'. The following excerpts describe important considerations within the standard when designing snow or other long duration loads on glass assemblies.

4.4.1 E1300 Section 1: scope

E1300 Section 1.2 limits practice to vertical and sloped glass resistance under wind, snow, and self-weight combined maximum loading of 10.1 kPa (210 psf); (this may be increased to 14.4 kPa (300 psf) in future revisions). Among other applications, the standard excludes the design of glass floor panels. Section 1.3 describes type and configuration limitations, including rectangular panel shape, continuous pin-type supports at two, three, or four edges, singular monolithic or laminated glass lites combined in an insulating glass (IG) unit. Section 1.4 excludes wired, patterned, etched, sandblasted, drilled, notched, or grooved glass with edge treatments that alter glass strength (this may be revised in the future to accommodate wired, patterned, etched, and sandblasted glass).

4.4.2 E1300 Section 5: significance and use

Refer to E1300 Section 5 for practice significance and use including important considerations for safe design; it states, 'considerations set forth in building codes along with criteria presented in safety glazing standards and site specific concerns may control the ultimate glass type and thickness selection'. In addition, the following must be considered when designing glass. Some of these factors are discussed in other chapters:

- Thermal stresses
- Spontaneous breakage of tempered glass
- Effects of wind-borne debris
- Behavior of glass fragments after breakage
- Seismic effects
- Heat flow
- Edge bite
- Noise abatement

Section 5.4 states, 'for situations not specifically addressed in this standard, the design professional shall use engineering analysis and judgment to determine the load resistance of glass in buildings'.

4.4.3 E1300 Section 6: procedure

E1300 Section 6 describes procedures for determining glass resistance to unfactored loads in various configurations and assemblies, and states, 'If the load resistance thus determined is less than the specified design load and duration, the selected glass types and thicknesses are not acceptable. If the load resistance is greater than or equal to the specified design load, then the glass types and thicknesses are acceptable for a probability of less than, or equal to, 8 in 1000.' The designer should note that combinations of load cases with differing durations require load duration factors (LDFs), as described in E1300 Appendix discussed below, to determine equivalent short duration load. Section 6 procedures include the following configurations:

- Monolithic single glazing simply supported continuously along four, three, or two (opposite) sides, or continuously supported along one edge (cantilever)
- Single-glazed laminated glass (LG) constructed with an interlayer simply supported continuously along four, three or two (opposite) sides, or continuously supported along one edge (cantilever) where in-service LG temperatures do not exceed 50 °C (122 °F)
- Insulating glass (IG) with glass lites of equal (symmetric) or different

(asymmetric) glass type and thickness simply supported continuously along four sides, including:

— IG with one monolithic lite and one laminated lite under short duration load
— IG with laminated glass over laminated glass under short duration load
— IG with one monolithic lite and one laminated lite under long duration load
— IG with laminated glass over laminated glass under long duration load

4.4.4 E1300 Annexes to standard practice

E1300 Annexes include design information mandatory for use in standard practice. Annex A1 includes nonfactored load and deflection charts used in design procedures. Nonfactored load charts are based on the surface strength of weathered glass associated with in-service conditions for approximately 20 years. Annex A2 provides three design procedure examples using the nonfactored load charts to determine glass resistance and one example (presented separately in SI and Imperial units) determining the approximate center-of-glass deflection.

4.4.5 E1300 Appendices

ASTM E1300 Appendices provide voluntary procedures and background information for determining glass load resistance and deflection. Appendix X1 presents an optional procedure for calculating the approximate center-of-glass deflection of a monolithic rectangular glass plate with four edges simply supported, subjected to a uniform lateral load. Appendix X1 notes conditions for procedures when calculating deflection of laminated glass and insulating glass assemblies. Appendix X2 presents an alternate method for calculating the approximate center-of-glass deflection. Appendix X3 presents an optional procedure to determine alternate breakage probabilities for annealed glass plates under uniform loading. Appendix X4 describes the background to glass type factors and IG factors in standard practice tables, stating that rigorous engineering analysis accounting for geometrical nonlinearity (i.e. caused by large deflections), surface condition, surface compression, prestress area, support conditions, load type and duration, and other relevant parameters can result in type factors other than those tabulated. Appendix X5 describes the determination of IG load share factors as proportional to the stiffness of the glass lites. Appendix X5 states:

• 'Under short duration loads laminated glass is assumed to behave in a monolithic-like manner. The glass thickness used for calculating load sharing factors is the sum of the thickness of glass of the two plies (in accordance with Table 4).'

• 'Under long duration loads laminated glass is assumed to behave in a layered manner. The load sharing is then based on the individual ply thicknesses of the laminated glass.'

Appendix X6 provides load duration factors (LDF) applied to annealed glass load resistance for durations greater than 3 seconds. Appendix X7 presents an approximate technique to determine a design load that represents the combined effects of multiple loads of different duration, providing a method to combine load cases of different durations into an equivalent short duration (3 seconds) load combination. Appendix X8 presents a technique for estimating the maximum allowable surface stress associated with glass lites continuously supported along all edges, for use in independent stress analysis. The information is proposed for rigorous engineering analysis of special glass shapes with loads not covered in the standard procedures. Appendix X8 provides an equation for calculating glass allowable surface stress based on glass area, load duration, and probability of breakage. Appendix X9 provides a conservative estimate of maximum allowable edge stresses for glass lites associated with a maximum probability of breakage less than or equal to 8 in 1000 for a 3 second load duration. Stresses in Table X9.1 are proposed for use in rigorous engineering analysis, accounting for large deflections when required. Appendix X10 presents the basis of laminated glass behavior used in the standard practice and allows the use of alternate interlayer materials with Young's modulus (determined per ASTM D 4065) greater than or equal to 1.5 MPa (218 psi) at 50 °C (122 °F) under an equivalent 3 second load. Appendix X10 covers monolithic interlayers at least 0.38 mm (0.015 in) thick without multiple polymer layers. See Section 7 for Appendix X11 (expected in forthcoming E1300 update) for the formulation of laminated glass effective thickness for calculating glass stresses and deflections subject to uniform loads.

4.5   E1300 standard practice examples

In addition to Annex A2 examples, the following 10 examples describe ASTM E1300 standard practice procedures for determining glass load resistance and maximum deflection, assuming a probability of breakage of 8 in 1000 for rectangular geometries (note Example 4.5.8 demonstrates the method to determine probability of breakage of 1 in 1000). Designers may rework examples in combination for alternate glass assemblies, dimensions, loads, and probabilities of breakage.

4.5.1   Example with four-side supported monolithic panel

Determine the load resistance and center-of-glass deflection associated with a monolithic glass panel described below. Compare the load resistance to an applied long duration self-weight load only (no snow load). Assume the following:

— Proposed nominal thickness 4 mm (5/32 in) monolithic annealed (AN) panel, horizontally oriented (0° slope). There are screens installed below to prevent falling broken glass.

— Glass dimensions between continuous supports along four sides: 1600 mm × 1000 mm (63.0 in × 39.4 in).

• Determine the load resistance (LR) and applied load:

— Determine the nonfactored load (NFL) from E1300 Annex 1, Figure A1.4 upper chart (Fig. 4.4). Enter the horizontal axis of the nonfactored load chart at 1600 mm (63 in) and project a vertical line. Enter the vertical axis of the nonfactored load chart at 1000 mm (39.4 in) and project a horizontal line. Sketch a line of constant aspect ratio through the intersection of the lines and interpolate along this line to determine the nonfactored load. The nonfactored load is approximately 1.73 kPa (36.1 psf).

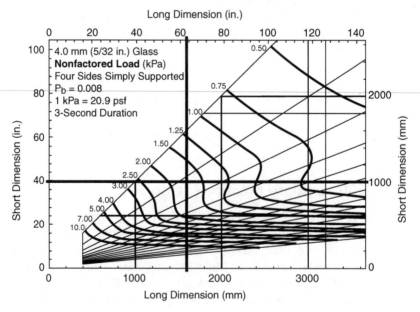

4.4 ASTM E1300, Annex A1, Figure A1.4 (upper chart).

Table 4.1   ASTM E1300, Table 1

| | GTF | |
Glass Type	Short Duration Load (3 sec)	Long Duration Load (30 day)
AN	1.0	0.43
HS	2.0	1.3
FT	4.0	3.0

— Determine the factored glass load resistance to uniform load through the product of the nonfactored load and the glass type factor for long duration load.
— Determine the glass type factor (GTF) for long duration load (self-weight) in E1300, Table 1 (Table 4.1). For long duration load the GTF for AN glass is 0.5. The factored glass resistance load is approximately 1.73 kPa × 0.5 = 0.865 kPa (18.1) psf).
— Compare the load resistance with applied load. The glass panel self-weight is the product of the nominal thickness 4 mm (5/32 in) and the material density $2500 \, kg/m^3 (0.090 \, lb/in^3)$. The self-weight uniform load is $4 \, mm/1000 \times 2500 \, kg/m^3 \times 9.81 \, m/s^2/1000 = 0.0981 \, kPa$ (2.05 psf). This is less than the factored glass resistance load of 0.865 kPa (18.1 psf). As the applied load is less than the load resistance, the proposed glass is acceptable.

• Determine the center-of-glass deflection from E1300, Figure A1.4, lower chart (Fig. 4.5):

— Calculate the panel aspect ratio:
  AR = 1600 mm/1000 mm (63.0 in/39.4 in) = 1.6.
— Calculate the panel area:
  area = 1.6 m × 1.0 m = $1.6 \, m^2 (2482 \, in^2 = 17.2 \, ft^2)$.
— Compute (load × $area^2$) as follows:
  $0.0981 \, kPa \times (1.6 \, m^2)^2 = 0.25 \, kN \, m^2 (0.606 \, kip \, ft^2)$.
— Project a vertical line upward from $0.25 \, kN \, m^2 (0.606 \, kip \, ft^2)$ along the lower (upper) horizontal axis.
— Project a horizontal line from the intersection point of the vertical line and an interpolated point between AR1 and AR2 to the right vertical axis and read the approximate center-of-glass deflection = 2 mm (0.08 in).

• Recheck the approximate center-of-glass deflection according to Appendix X2, where a = long dimension and b = short dimension, E = elastic modulus = $71.7 \times 10^6 \, kPa$ (10.4×10^6 psi), q = uniform lateral load, t = actual plate thickness:

— a/b = 1600 mm/1000 mm (63.0 in/39.4 in) = 1.6

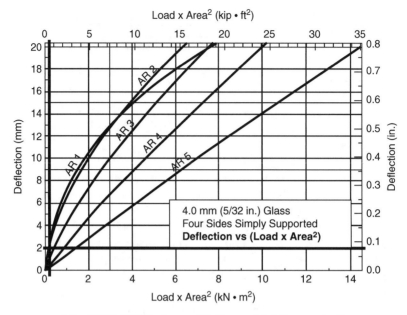

4.5 ASTM E1300, Annex A1, Figure A1.4 (lower chart).

— $t = 3.78\,\text{mm}\ (0.149\,\text{in})$ from Table 4 (Table 4.2)
— $r_0 = 0.553 - 3.83(a/b) + 1.11(a/b)^2 - 0.0969(a/b)^3 = -3.13$
  (Equation X2.2)
— $r_1 = -2.29 + 5.83(a/b) - 2.17(a/b)^2 - 0.2067(a/b)^3 = 2.33$ (Equation X2.3)
— $r_2 = 1.485 - 1.908(a/b) + 0.815(a/b)^2 - 0.0822(a/b)^3 = 0.18$
  (Equation X2.4)
— $x = \ln\{\ln[q(ab)^2/Et^4]\}$ (Equation X2.5)
  $= \ln\{\ln[(0.25)(63.0 \times 39.4)^2/(10.4 \times 10^6)(0.149)^4]\} = 1.74$
— $w = t \times \exp(r_0 + r_1 \times x + r_2 \times x^2)$ (Equation X2.1)
  $= 0.149 \times \exp(-3.13 + 2.33 \times 1.74 + 0.18 \times 1.74^2)$
  $= 0.65\,\text{mm}\ (0.03\,\text{in})$

This deflection is lower than the deflection determined in Fig. 4.5 above.

Table 4.2 ASTM E1300, Table 4

Nominal Thickness or Designation, mm (in.)	Minimum Thickness, mm (in.)
2.5 (3/32)	2.16 (0.085)
2.7 (lami)	2.59 (0.102)
3.0 (1/8)	2.92 (0.115)
4.0 (5/32)	3.78 (0.149)
5.0 (3/16)	4.57 (0.180)
6.0 (1/4)	5.56 (0.219)
8.0 (5/16)	7.42 (0.292)
10.0 (3/8)	9.02 (0.355)
12.0 (1/2)	11.91 (0.469)
16.0 (5/8)	15.09 (0.595)
19.0 (3/4)	18.26 (0.719)
22.0 (7/8)	21.44 (0.844)

4.5.2 Example with three-side supported monolithic panel

Determine the load resistance and maximum glass deflection associated with a monolithic glass panel described below. Compare the load resistance to an applied long duration self-weight and a given snow load of 2.2 kPa (46 psf). Identify the location of the maximum deflection. Assume the following:

— Proposed nominal thickness 10 mm (3/8 in) heat strengthened (HS) panel at 1% slope, or 10 mm rise per 1 m run (1/8 in per ft) toward the unsupported edge. There are screens installed below to prevent falling broken glass.
— Glass dimensions with continuous supports along three sides: 1500 mm (59 in) (supported edges) × 1000 mm (39.4 in) (unsupported edge).

• Determine and compare load resistance (LR) and applied load:

— NFL (nonfactored load): from E1300 Figure A1.20, upper chart, NFL = 2.2 kPa (46 psf) (Fig. 4.6).
— GTF (glass type factor): from E1300, Table 1, GTF = 1.3 for HS long duration loads (Table 4.1).
— LR (load resistance): LR = NFL × GTF = 2.2 × 1.3 = 2.9 kPa (59.7 psf).
— SW (self-weight) = $10\,mm/1000 \times 2500\,kg/m^3 \times 9.81\,m/s^2/1000$ = 0.24 kPa (5.1 psf); SL (snow loads) = 2.2 kPa (46 psf); applied load = combined SL + SW = 2.2 + 0.24 = 2.4 kPa (51 psf).
— Compare the applied load with LR: 2.44 kPa (51 psf) < 2.9 kPa (59.7 psf). As the applied load is less than the load resistance, the proposed glass is acceptable.

• Determine the maximum glass deflection under applied load from

Length of Parallel Supported Edges (in.)

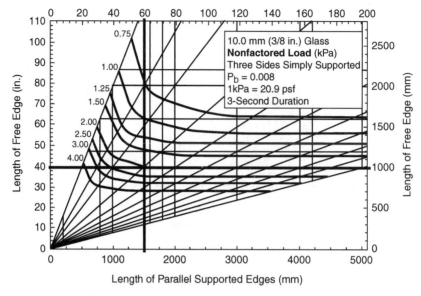

4.6 ASTM E1300, Annex A1, Figure A1.20 (upper chart).

Load x L⁴ (kip • ft²) [L Denotes Length of Free Edge]

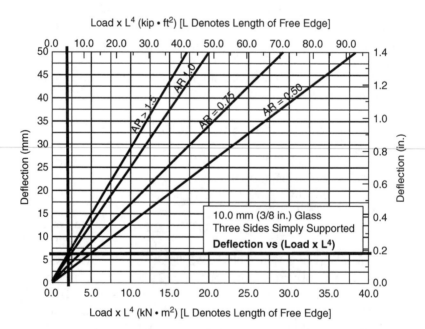

Load x L⁴ (kN • m²) [L Denotes Length of Free Edge]

4.7 ASTM E1300, Annex A1, Figure A1.20 (lower chart).

E1300, Figure A1.20, lower chart (Fig. 4.7):

— Aspect ratio = 1500 mm/1000 mm (59 in/39.4 in) = 1.5.
— Compute (load × L^4) as follows, where L = free edge length:
  2.44 kPa × $(1.0\,\text{m})^4$ = 2.44 kN m^2(5.9 kip ft^2).
— Project a vertical line from 2.44 kN m^2 (5.9 kip ft^2) along the horizontal axis.
— Project a horizontal line from the intersection point of the vertical line and an interpolated point along AR > 1.5 to the vertical axis and read the approximate unsupported glass edge deflection = 6 mm (0.23 in). The panel will drain snow and water from the unsupported edge.

4.5.3 Example with two-side supported (parallel edges) monolithic panel

Determine the load resistance and center-of-glass deflection associated with a monolithic glass panel described below. Compare the load resistance to an applied long duration self-weight and a given snow load of 1.44 kPa (30 psf). Comment on deflection serviceability considerations. Assume the following:

— Proposed nominal thickness 12 mm (1/2 in) monolithic fully tempered (FT) panel at 1% slope, or 10 mm rise per 1 m run (1/8 in per ft) toward the unsupported edge. There are screens installed below to prevent falling broken glass.
— Glass dimensions with continuous supports along two parallel edges: 914 mm (36 in) (supported edges) × 1828 mm (72 in) (unsupported edges).

• Determine the load resistance (LR) and applied load:

— NFL (nonfactored load): from E1300, Figure A1.25 (upper chart), NFL = 0.96 kPa (20 psf) (Fig. 4.8).
— GTF (glass type factor): from E1300, Table 1, GTF = 3.0 for FT long duration loads (Table 4.1).
— LR (load resistance): LR = NFL × GTF = 0.96 × 3.0 = 2.88 kPa (60 psf).
— SW (self-weight) = 12 mm/1000 × 2500 kg/m^3 × 9.81 m/s^2/1000 = 0.29 kPa (6.06 psf); SL (snow loads) = 1.44 kPa (30 psf); applied load = combined SL + SW = 1.44 + 0.29 = 1.73 kPa (36 psf).
— Compare applied load with LR: 1.73 kPa (36 psf) < 2.88 kPa (60 psf). As the applied load is less than load resistance, the proposed glass is acceptable.
• Determine the approximate maximum glass deflection from the lower

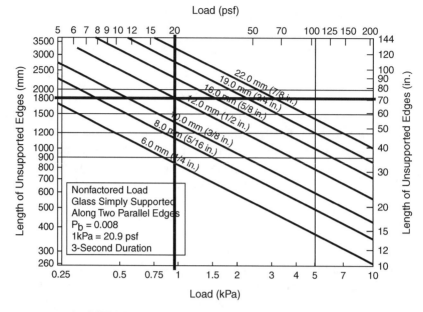

4.8 ASTM E1300, Annex A1, Figure A1.25 (upper chart).

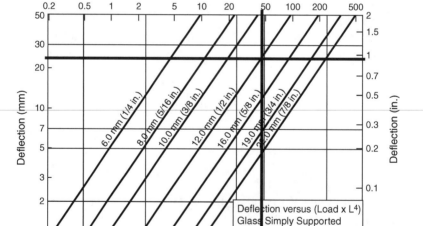

4.9 ASTM E1300, Annex A1, Figure A1.25 (lower chart).

chart of ASTM E1300, Figure A1.25 (Fig. 4.9):

— Compute (load $\times L^4$) as follows, where L = free edge length: $1.73\,\text{kPa} \times (1.83\,\text{m})^4 = 19.40\,\text{kN m}^2(46.9\,\text{kip ft}^2)$.
— Project a vertical line from $19.40\,\text{kN m}^2(46.9\,\text{kip ft}^2)$ along the horizontal axis.
— Project a horizontal line from the intersection point of the vertical line and an interpolated point along thickness = 12 mm to the vertical axis and read the approximate glass edge deflection = 23 mm (0.875 in).

Many glass suppliers limit the maximum center-of-glass deflection to the glass span divided by 100, or $\delta_{max} = L/100$. For the example above, this translates into a maximum deflection of 18.3 mm (0.72 in). The deflection due to the applied load exceeds $\delta_{max} = L/100$. In addition, the design may require smaller deflections for acceptable visual appearance. For these reasons, consider using a thicker glass in order to lower service deflections. Determine the maximum glass deflection using 19 mm (3/4 in) thick fully tempered (FT) monolithic glass.

• Determine the approximate maximum glass deflection from the lower chart of ASTM E1300, Figure A1.25 (Fig. 4.10):
— Project a vertical line from $19.40\,\text{kN m}^2(46.9\,\text{kip ft}^2)$ along the horizontal axis.

Load x L⁴ (kip • ft²) [L Denotes Length of Unsupported Edges]

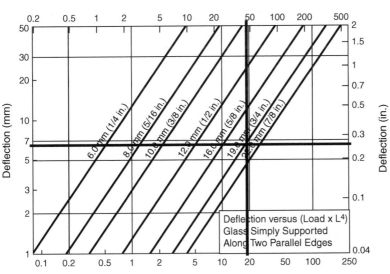

Load x L⁴ (kN • m²) [L Denotes Length of Unsupported Edges]

4.10 ASTM E1300, Annex A1, Figure A1.25 (lower chart).

— Project a horizontal line from the intersection point of the vertical line
and an interpolated point along thickness = 19 mm to the vertical axis
and read the approximate glass edge deflection = 6.5 mm (0.256 in).
The deflection ratio = 1830 mm/6.5 mm (72 in/0.256 in) = 282 > 100.
The service deflection under the applied load is acceptable.

4.5.4 Example with four-side supported laminated panel

- Determine the load resistance and center-of-glass deflection using
standard practice. In addition, determine the required panel slope to
avoid ponding associated with a laminated glass panel described below.
Compare the load resistance to an applied long duration self-weight and
a given snow load of 2.4 kPa (50 psf). Assume the following:

— The proposed assembly including nominal thickness 6 mm (1/4 in) heat
strengthened (HS) panel plus 1.5 mm (0.06 in) PVB interlayer plus
nominal thickness 6 mm (1/4 in) heat strengthened (HS) panel at 2%
slope, or 21 mm rise per 1 m run (1/4 in per ft).
— Glass dimensions between continuous supports along four sides: 1829
mm × 1829 mm (72 in × 72 in).
— In-service laminated glass temperatures do not exceed 50 °C (122°F).

- Determine and compare load resistance (LR) and applied load:

— NFL (nonfactored load) from the upper chart of ASTM E1300, Figure
A1.31 = 3.4 kPa (71.1 psf) (Fig. 4.11).
— GTF (glass type factor) from Table 1 = 1.3 for HS long duration loads
(Table 4.1).
— LR (load resistance): LR = NFL × GTF = 3.4 kPa × 1.3 = 4.42 kPa
(92.4 psf).
— SW (self-weight) = 12 mm/1000 × 2500 kg/m^3 × 9.81 m/s^2/1000
= 0.29 kPa (6.06 psf); SL = 2.4 kPa (50 psf); total load = SL + SW
= 2.4 + 0.29 = 2.69 kPa (56.2 psf).
— Compare the applied loads to the load resistance (LR): 2.69 kPa (56.2
psf) < 4.42 kPa (92.4 psf). As the applied load is less than load
resistance, the proposed glass is acceptable.

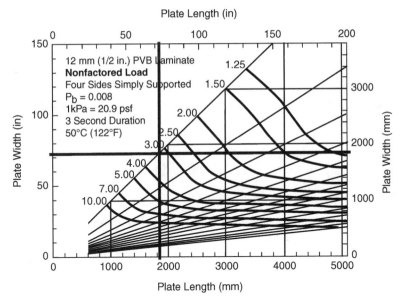

4.11 ASTM E1300, Annex A1, Figure A1.31 (upper chart).

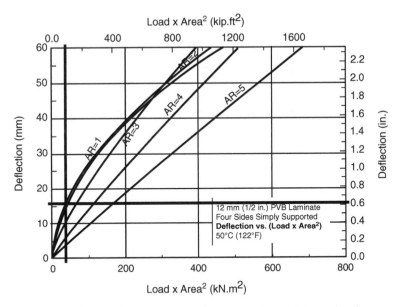

4.12 ASTM E1300, Annex A1, Figure A1.31 (lower chart).

- Determine the maximum glass deflection under the applied load from the lower chart of ASTM E1300 Annex 1, Figure A1.31 (Fig. 4.12):

— Aspect ratio = 1829 mm/1829 mm (72 in/72 in) = 1.0.
— Calculate the glass area as follows:
   area = 1.83 m × 1.83 m = 3.35 m^2 (5169 in^2 = 35.9 ft^2).
— Compute (load × area^2) as follows:
   2.69 kPa × (3.35 m^2)^2 = 30.2 kN m^2 (73.1 kip ft^2).
— Project a vertical line from 30.2 kN m^2 (73.1 kip ft^2) along the horizontal axis.
— Project a horizontal line from the intersection point of the vertical line and AR = 1 to the vertical axis and read the approximate center-of-glass deflection = 16 mm (0.63 in). The deflection ratio = 1829 mm/16 mm = 114.

• In addition to standard practice, determine the required slope for the assembly to avoid ponding. Confirm that the assembly slope is greater than or equal to the maximum slope of the sheet's deflected shape (θ_{max}) (Fig. 4.13):

— Maximum glass deflection, δ = 16 mm (0.63 in). Assume the deflected shape is parabolic (conservative) with length L and maximum deflection δ.
— The vertical displacement in the downward direction (y) at a distance measured from the side of the sheet (x) can be determined from $y = m \times (x - L/2)^2 + \delta$.
— At the supports, $y = x = 0$. Therefore, $m = 4\delta/L^2$ and $y = (4\delta/L^2) \times (x - L/2)^2 + \delta$.
— Taking the first derivative of y with respect to x yields: $dy/dx = (8\delta/L^2) \times (x - L/2)$.
— $\tan(\theta) = dy/dx$. For small values of θ, $\theta = \tan(\theta) = dy/dx = (8\delta/L^2) \times (x - L/2)$.

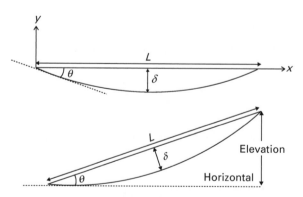

4.13 Confirm assembly slope is greater than or equal to the maximum slope of deflected shape (θ_{max}).

— θ_{max} occurs at the supports, $x = 0$ or L; $\theta_{max} = (8\delta/L^2)$
  $\times(L - L/2) = 4\delta/L$.
— Converting to degrees: $\theta_{max} = (180°/\pi) \times (4\delta/L) = (720\delta)/(\pi L)$.
— Substituting values for δ and L: $\theta_{max} = (720 \times 16)/(\pi \times 1829) = 2.0°$.
  Therefore the required slope for the assembly is $2.0°$.
— Given that tan $(2.0°) \times 1829 = 64.0$ mm (2.52 in), one end of the
  assembly must be elevated at least 64.0 mm (2.52 in) from the other end.
— Required assembly slope $= 64.0/1829 = 0.035$, or 35 mm/m (13/32
  in/ft) min.

4.5.5 Example with four-side supported IG unit

- Determine the load resistance (LR) and center-of-glass deflection
  associated with an IG unit described below. Compare the load resistance
  to an applied self-weight (SW), a given snow load (SL) of 1.0 kPa (21
  psf), and a given wind load (WL) of 1.5 kPa (31.3 psf). Assume the
  following:

— The proposed assembly includes nominal thickness 6 mm (0.25 in) heat
  strengthened (HS) monolithic outer panel, an airspace of 12 mm (0.5
  in), and a nominal thickness 10 mm (0.375 in) heat strengthened (HS)
  laminated inner panel.
— Glass dimensions: 914 mm × 1524 mm (36 in × 60 in).
— Winter temperatures are below 50°C (122°F); summer temperatures
  exceed 50 °C (122 °F); for high temperatures where interlayer stiffness is
  significantly decreased, treat laminated assemblies as noncomposite
  layered assemblies.

- Determine the nonfactored load and the load resistance:

— Determine NFL (nonfactored load) from the upper charts of ASTM
  E1300 Annex 1, Figures A1.6 and A.1.8: NFL1 = 2.73 kPa (57.1 psf)
  (Fig. 4.14); NFL2 = 5.20 kPa (108.6 psf) (Fig. 4.15).
— Determine the GTF (glass type factor) from Tables 2 and 3; from Table
  2 (short duration), GTF1 = 1.8 and GTF2 = 1.8 (Table 4.3); from
  Table 3 (long duration), GTF1 = 1.25 and GTF2 = 1.25 (Table 4.4).
— Determine the LS (load share factor): from Table 5, LS1 = 5.26 and
  LS2 = 1.23 (short duration) (Table 4.5) and from Table 6, LS1 = 2.11
  and LS2 = 1.90 (long duration) (Table 4.6).

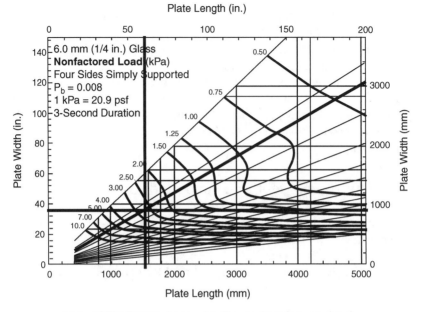

4.14 ASTM E1300, Annex A1, Figure A1.6 (upper chart).

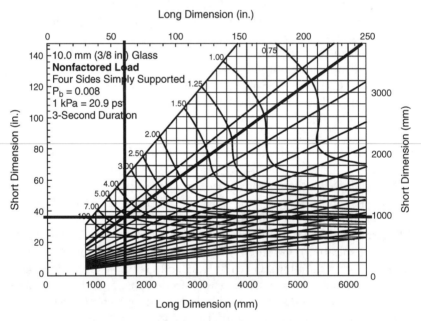

4.15 ASTM E1300, Annex A1, Figure A1.8 (upper chart).

Table 4.3 ASTM E1300, Table 2 (short duration)

Lite No. 1 Monolithic Glass or Laminated Glass Type	Lite No. 2 Monolithic Glass or Laminated Glass Type					
	AN		HS		FT	
	GTF1	GTF2	GTF1	GTF2	GTF1	GTF2
AN	0.9	0.9	1.0	1.9	1.0	3.8
HS	1.9	1.0	1.8	1.8	1.9	3.8
FT	3.8	1.0	3.8	1.9	3.6	3.6

Table 4.4 ASTM E1300, Table 3 (long duration)

Lite No. 1 Monolithic Glass or Laminated Glass Type	Lite No. 2 Monolithic Glass or Laminated Glass Type					
	AN		HS		FT	
	GTF1	GTF2	GTF1	GTF2	GTF1	GTF2
AN	0.39	0.39	0.43	1.25	0.43	2.85
HS	1.25	0.43	1.25	1.25	1.25	2.85
FT	2.85	0.43	2.85	1.25	2.85	2.85

- Determine LR (load resistance):
  $LR1_S = NFL1 \times GTF1 \times LS1 = 2.73 \times 1.8 \times 5.26 = 25.8\,kPa$ (540 psf)
  (short duration),
  $LR2_S = NFL2 \times GTF2 \times LS2 = 5.2 \times 1.8 \times 1.23 = 11.5\,kPa$ (241 kPa)
  (short duration);
  $LR1_L = NFL1 \times GTF1 \times LS1 = 2.73 \times 1.25 \times 2.11 = 7.20\,kPa$
  (375 psf) (long duration),
  $LR2_L = NFL2 \times GTF2 \times LS2 = 5.2 \times 1.25 \times 1.90 = 12.35\,kPa$
  (167 psf) (long duration),
- The lowest load resistance value control, therefore, is
  $LR1_L = 7.20\,kPa$ (150 psf).

- Determine the applied loads: SW load = 16 mm/1000 × 2500 kg/m^3 × 0.00981 kN/kg = 0.39 kPa (8.1 psf); SL = 1.0 kPa (21 psf); WL = 1.5 kPa (31.3 psf); total load = 0.39 + 1.0 + 1.5 kPa = 2.89 kPa (60.4 psf).
- Compare the applied loads to the load resistance (LR): 2.89 kPa ≤ 7.20 kPa. The calculated load due to self-weight, snow load, and wind load is less than the load resistance. Therefore, the glass thickness is acceptable for the applied loads.
- Determine the load share on each lite per Appendix X5 and inverting Equations X5.1 and X5.2: for short duration and long duration loading (LS factors are the same in both cases), Lite1 carries $[6^3/(6^3 + 10^3)] = 18\%$ of the load share and Lite2 carries $[10^3/(6^3 + 10^3)] = 82\%$ of the load share. Note, from Equation X5.3, that these values assume that the laminated glass panel behaves in a monolithic-like manner under short duration loads and in a layered manner under long duration loads.
- Recheck LS from Equation X5.3 and LR for layered behavior of lower laminated glass panel under long duration loads:

Table 4.5 ASTM E1300, Table 5 (short duration)

NOTE 1—Lite No. 1 Monolithic glass, Lite No. 2 Monolithic glass, short or long duration load, or Lite No. 1 Monolithic glass, Lite No. 2 Laminated glass, short duration load only, or Lite No. 1 Laminated Glass, Lite No. 2 Laminated Glass, short or long duration load.

		Lite No. 2																				
		Monolithic Glass				Monolithic Glass, Short or Long Duration Load or Laminated Glass, Short Duration Load Only																
Nominal Thickness		2.5 (3/32)		2.7 (lami)		3 (1/8)		4 (5/32)		5 (3/16)		6 (1/4)		8 (5/16)		10 (3/8)		12 (1/2)		16 (5/8)		19 (3/4)	
mm	(in.)	LS1	LS2	LS1	LS2	LS1	LS2	LS1	LS2	LS1	LS2	LS1	LS2	LS1	LS2	LS1	LS2	LS1	LS2	LS1	LS2	LS1	LS2
2.5	(3/32)	2.00	2.00	2.73	1.58	3.48	1.40	6.39	1.19	10.5	1.11	18.1	1.06	41.5	1.02	73.8	1.01	169.	1.01	344.	1.00	606.	1.00
2.7	(lami)	1.58	2.73	2.00	2.00	1.70	2.00	4.12	1.32	6.50	1.18	10.9	1.10	24.5	1.04	43.2	1.02	98.2	1.01	199.	1.01	351.	1.00
3	(1/8)	1.40	3.48	1.70	2.43	2.00	2.00	3.18	1.46	4.83	1.26	7.91	1.14	17.4	1.06	30.4	1.03	68.8	1.01	140.	1.01	245.	1.00
4	(5/32)	1.19	6.39	1.32	4.12	1.46	3.18	2.00	2.00	2.76	1.57	4.18	1.31	8.53	1.13	14.5	1.07	32.2	1.03	64.7	1.02	113.	1.01
5	(3/16)	1.11	10.5	1.18	6.50	1.26	4.83	1.57	2.76	2.00	2.00	2.80	1.56	5.27	1.23	8.67	1.13	18.7	1.06	37.1	1.03	64.7	1.02
6	(1/4)	1.06	18.1	1.10	10.9	1.14	7.91	1.31	4.18	1.56	2.80	2.00	2.00	3.37	1.42	5.26	1.23	10.8	1.10	21.1	1.05	36.4	1.03
8	(5/16)	1.02	41.5	1.04	24.5	1.06	17.4	1.13	8.53	1.23	5.27	1.42	3.37	2.00	2.00	2.80	1.56	5.14	1.24	9.46	1.12	15.9	1.07
10	(3/8)	1.01	73.8	1.02	43.2	1.03	30.4	1.07	14.5	1.13	8.67	1.23	5.26	1.56	2.80	2.00	2.00	3.31	1.43	5.71	1.21	9.31	1.12
12	(1/2)	1.01	169.	1.01	98.2	1.01	68.8	1.03	32.2	1.06	18.7	1.10	10.8	1.24	5.14	1.43	3.31	2.00	2.00	3.04	1.49	4.60	1.28
16	(5/8)	1.00	344.	1.01	199.	1.01	140.	1.02	64.7	1.03	37.1	1.05	21.1	1.12	9.46	1.21	5.71	1.49	3.04	2.00	2.00	2.76	1.57
19	(3/4)	1.00	606.	1.00	351.	1.00	245.	1.01	113.	1.02	64.7	1.03	36.4	1.07	15.9	1.12	9.31	1.28	4.60	1.57	2.76	2.00	2.00

Lite No. 1 — Monolithic Glass

Table 4.6 ASTM E1300, Table 6 (long duration)

NOTE 1—Lite No. 1 Monolithic glass, Lite No. 2 Laminated glass, long duration load only.

Lite No. 1 Monolithic Glass Nominal Thickness		Lite No. 2 Laminated Glass												
		5 (3/16)		6 (1/4)		8 (5/16)		10 (3/8)		12 (1/2)		16 (5/8)		19 (3/4)	
mm	(in.)	LS1	LS2	LS1	LS2	LS1	LS2	LS1	LS2	LS1	LS2	LS1	LS2	LS1	LS2
2.5	(3/32)	3.00	1.50	4.45	1.29	11.8	1.09	20.0	1.05	35.2	1.03	82.1	1.01	147	1.01
2.7	(lam)	2.16	1.86	3.00	1.50	7.24	1.16	12.0	1.09	20.8	1.05	48.0	1.02	85.5	1.01
3	(1/8)	1.81	2.24	2.39	1.72	5.35	1.23	8.68	1.13	14.8	1.07	33.8	1.03	60.0	1.02
4	(5/32)	1.37	3.69	1.64	2.56	3.00	1.50	4.53	1.28	7.34	1.16	16.1	1.07	28.1	1.04
5	(3/16)	1.21	5.75	1.36	3.75	2.13	1.88	3.00	1.50	4.60	1.28	9.54	1.12	16.4	1.07
6	(1/4)	1.12	9.55	1.20	5.96	1.63	2.59	2.11	1.90	3.00	1.50	5.74	1.21	9.54	1.12
8	(5/16)	1.05	21.3	1.09	12.8	1.27	4.76	1.47	3.13	1.84	2.19	3.00	1.50	4.60	1.28
10	(3/8)	1.03	37.4	1.05	22.1	1.15	7.76	1.26	4.83	1.47	3.13	2.11	1.90	3.00	1.50
12	(1/2)	1.01	85.0	1.02	49.7	1.06	16.6	1.11	9.84	1.20	5.92	1.48	3.07	1.87	2.15
16	(5/8)	1.01	172	1.01	100	1.03	32.8	1.06	19.0	1.10	11.0	1.24	5.23	1.43	3.35
19	(3/4)	1.00	304	1.01	176	1.02	57.2	1.03	32.8	1.06	18.7	1.13	8.46	1.24	5.15
22	(7/8)	1.00	440	1.00	256	1.01	82.5	1.02	47.2	1.04	26.7	1.09	11.8	1.17	7.02

LS2 = $[5^3/6^3 + 2 \times 5^3)] = 27\%$ (inverted LS2 = 3.73) and LS1 = $[6^3/(6^3 + 2 \times 5^3)] = 46\%$ (inverted LS1 = 2.16);
LR1$_L$ = NFL1 × GTF1 × LS1 = 2.73 × 1.25 × 2.16 = 7.37 kPa (153 psf) (long duration),
LR2$_L$ = NFL2 × GTF2 × LS2 = 5.2 × 1.25 × 3.73 = 24.2 kPa (505 psf) (long duration).

- Recompare the applied loads to the load resistance (LR): 2.89 kPa ≤ 7.37 kPa. The calculated load due to self-weight, snow load, and wind load (included conservatively) is less than the load resistance. Therefore, the glass thickness is still acceptable for the applied loads.
- Determine the approximate center-of-glass deflection for the upper and lower IG panels using E1300 Appendix 1. By Equation X1.1.2, for a lower laminated glass panel under long duration loads, consider the approximate deflection as the single lite deflection at half of the design load. Determine the deflection for 5 mm monolithic (single lite of 10 mm laminated assembly) under 0.5 × (total load = 2.89 kPa) = 1.45 kPa (30.2 psf).

— Calculate the aspect ratio (AR) = 1524/914 = 1.67; calculate the area squared $A^2 = (1524 \times 914)^2 = 1.94 \times 10^{12}$; $E = 71.7 \times 10^6$ kPa; t_{actual} = 4.57 mm (0.18 in) from Table 4 (Table 4.2); t^4 = 436 mm^4.
— Determine the natural log of the nondimensional lateral load (q^*), using Equation X1.3 in Appendix X1; $q^* = qA^2/Et^4 = 1.45 \times (1.94 \times 10^{12})/(71.7 \times 10^6 \times 436) = 90$, ln($q^*$) = 4.50.
— Project a vertical line from AR = 1.67 and a horizontal line from ln(q^*) = 4.50 on Figure X1.1 (Fig. 4.16) to determine the nondimensional

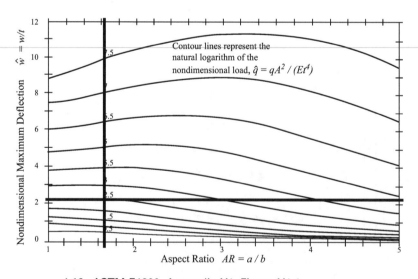

4.16 ASTM E1300, Appendix X1, Figure X1.1.

maximum deflection $\hat{w} = 2.25$ mm. Calculate the maximum deflection w from $\hat{w} = w/t_{actual}$: $w = 2.25 \times 4.57 = 10.6$ mm (0.42 in). This corresponds to a span–deflection ratio $= 914/10.6 = L/86$. Based on a typical span–deflection ratio of $L/100$ and potential for ponding, consider a thicker lower panel laminated glass assembly for long duration load deflections.

— Check the upper panel deflection per the procedures above and provide an adequate slope to avoid ponding.

4.5.6 Example with four-side supported insulating glass (IG) unit

● Determine the load resistance and center-of-glass deflection. In addition to standard practice, determine the required panel slope to avoid ponding, associated with an IG unit described below. Compare the load resistance to an applied long duration self-weight and a given snow load (SL) of 1.44 kPa (30 psf). Assume the following:

— The proposed assembly includes a nominal thickness 6 mm (0.25 in) thick fully tempered (FT) monolithic outer panel, plus an airspace of 16 mm (0.625 in), plus a nominal thickness 6 mm (0.25 in) thick annealed (AN) laminated inner panel.

— Glass dimensions: 1219 mm (48 in) $\times$ 2438 mm (96 in).

— The panel is sloped in the long dimension.

● Determine the nonfactored load and the load resistance:

— NFL (nonfactored load) from the upper charts of ASTM E1300 Annex 1, Figure A1.6 and Figure A.1.28: NFL1 $=$ 1.37 kPa (29 psf) (Fig. 4.17); NFL2 $=$ 1.47 kPa (31 psf) (Fig. 4.18).

— Determine the GTF (glass type factor) from ASTM E1300, Tables 2 and 3. From Table 2 (short duration), GTF1 $=$ 3.8 and GTF2 $=$ 1.0 (Table 4.3). From Table 3 (long duration), GTF1 $=$ 2.85 and GTF2 $=$ 0.5 (Table 4.4);

— Determine the LS (load share factor) from Tables 5 and 6. From Table 5, LS1 $=$ 2.0 and LS2 $=$ 2.0 (for short duration loading) (Table 4..5). From Table 6, LS1 $=$ 1.20 and LS2 $=$ 5.96 (for long duration loading) (Table 4.6)

— Determine LR (load resistance):
$LR1_L = NFL1 \times GTF1 \times LS1 = 1.37 \times 2.85 \times 1.20 = 4.7$ kPa (98 psf) (long duration),

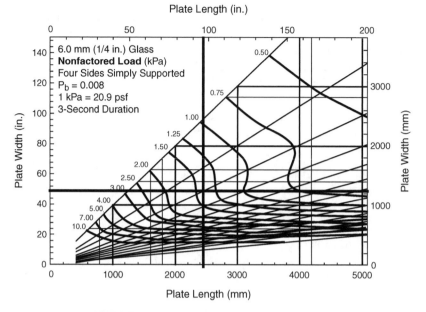

4.17 ASTM E1300, Annex A1, Figure A1.6 (upper chart).

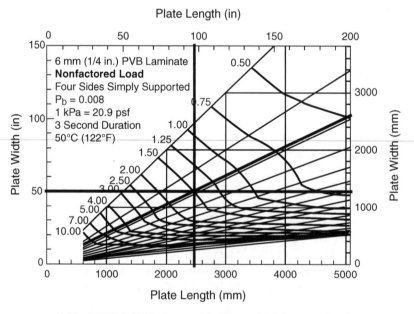

4.18 ASTM E1300, Annex A1, Figure A1.8 (upper chart).

$\text{LR2}_L = \text{NFL2} \times \text{GTF2} \times \text{LS2} = 1.47 \times 0.5 \times 5.96 = 4.4 \, \text{kPa}$ (92 psf)
(long duration);
$\text{LR1}_S = \text{NFL1} \times \text{GTF1} \times \text{LS1} = 1.37 \times 3.8 \times 2.0 = 10.4 \, \text{kPa}$ (217 psf)
(short duration),
$\text{LR2}_S = \text{NFL2} \times \text{GTF2} \times \text{LS2} = 1.47 \times 1.0 \times 2.0 = 2.9 \, \text{kPa}$ (61 psf)
(short duration).
— The lowest load resistance value controls:
$\text{LR} = \text{LR2}_S$ (short duration) $= 2.9 \, \text{kPa}$ (61 psf).

• Determine the applied loads:

— SW load $= 12 \, \text{mm}/1000 \times 2500 \, \text{kg/m}^3 \times 0.00981 \, \text{kN/kg} = 0.29 \, \text{kPa}$ (6.15 psf); SL $= 1.44 \, \text{kPa}$ (30 psf); total load $= \text{SL} + \text{SW} = 1.44 + 0.29 = 1.73 \, \text{kPa}$ (36.15 psf).
• Compare the applied loads to the load resistance (LR): 1.76 kPa $\leq$ 2.9 kPa. The calculated load due to self-weight and snow load is less than the load resistance. Therefore, the proposed IG panel is acceptable for applied loads.
• Determine the load share on each lite (see the LS values above):

— For short duration loading, Lite1 carries $[2/(2 + 2)] = 50\%$ of the load share and Lite2 carries $[2 / (2 + 2)] = 50\%$ of the load share.
— For long duration loading, Lite1 carries $[5.96/(1.2 + 5.96)] = 83\%$ of the load share and Lite2 carries $[1.2 / (1.2 + 5.96)] = 17\%$ of the load share.

• Determine the approximate center-of-glass deflection for each lite, using the lower chart of ASTM E1300 Annex 1, Figures A1.6 and A.1.28:

— For lite 1: load $\times$ area$^2 = 0.83 \times 1.73 \times (1.219 \times 2.438)^2 = 12.7 \, \text{KN m}^2$ (30.7 kip ft^2). From the lower chart (Fig. 4.19), $\delta_1 = 18.3 \, \text{mm}$ (0.72 in).
— For lite 2: load $\times$ area$^2 = 0.17 \times 1.73 \times (1.219 \times 2.438)^2 = 2.6 \, \text{KN m}^2$ (6.1 kip ft^2). From the lower chart (Fig. 4.20), $\delta_2 = 9.9 \, \text{mm}$ (0.39 in).

• In addition to standard practice, determine the required slope for the IG unit (see Example 4.5.4 procedure):

— Maximum deflection of the outer lite $\delta_1 = 18.3 \, \text{mm}$ (0.72 in).
— $\theta_{max} = (180°/\pi) \times (4\delta/L) = (720\delta)/(\pi L) = (720 \times 18.3)/(\pi \times 2438) = 1.72°$. Therefore the required slope for the assembly is 1.72°.
— Given that tan $(1.72°) \times 2438 = 73.2 \, \text{mm}$ (2.88 in), one end of the assembly must be elevated at least 73.2 mm (2.88 in).

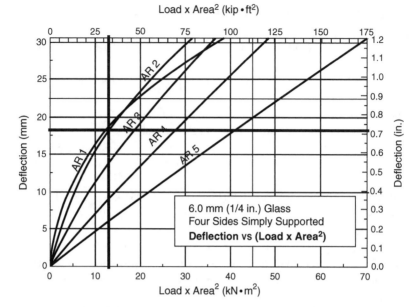

4.19 ASTM E1300, Annex A1, Figure A1.6 (lower chart).

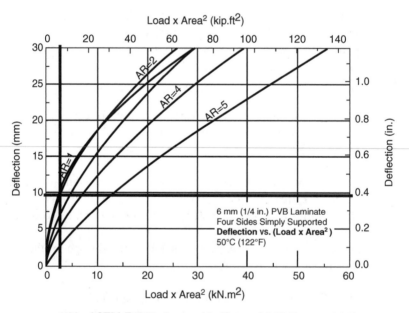

4.20 ASTM E1300, Annex A1, Figure A1.28 (lower chart).

— Required assembly slope = 72.0/2438 = 0.029 or 29 mm/m (3/8 in/ft) min.

4.5.7 Example with four-side supported insulating glass (IG) unit

Determine the nonfactored load, factored load resistance, and maximum deflection associated with an IG panel under self-weight of the glass (SW) and a given snow load (SL) of 1.0 kPa (21 psf). Assume the following:

— The proposed assembly includes an outer panel composed of a nominal thickness 8 mm (0.3125 in) thick fully tempered (FT) laminated outer panel, plus an airspace of 16 mm (0.625 in) plus a nominal 10 mm (0.375 in) thick annealed (AN) laminated inner panel.
— Panel dimensions: 914 mm (36 in) × 2438 mm (96 in) in size.

• Determine the load resistance:

— Determine NFL (nonfactored load) from the upper charts of ASTM E1300 Annex 1, Figures A1.6 and A.1.28: NFL1 = 1.6 kPa (33.4 psf) (Fig. 4.21); NFL2 = 1.75 kPa (36.6 psf) (Fig. 4.22).
— Determine the GTF (glass type factor) from ASTM E1300, Tables 2 and 3; from Table 2 (short duration), GTF1 = 3.8 and GTF2 = 1.0

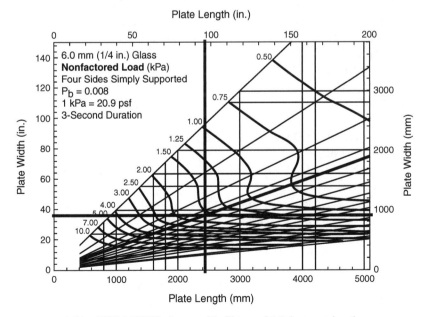

4.21 ASTM E1300, Annex A1, Figure A1.6 (upper chart).

Plate Length (in)

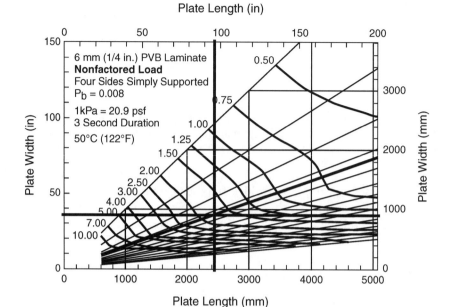

4.22 ASTM E1300, Annex A1, Figure A1.28 (upper chart).

(Table 4.3); from Table 3 (long duration), GTF1 = 2.85 and GTF2 = 0.5 (Table 4.4).

— Determine the LS (load share factor) from Tables 5 and 6; from Table 5, LS1 = 2.8 and LS2 = 1.56 (short duration) (Table 4.5); from Table 6, LS1 = 2.8 and LS2 = 1.56 (long duration) (Table 4.6).

— Determine LR (load resistance):
  $LR1_L = NFL1 \times GTF1 \times LS1 = 1.6 \times 2.85 \times 2.8 = 12.8\,kPa$ (267 psf) (long duration),
  $LR2_L = NFL2 \times GTF2 \times LS2 = 1.75 \times 0.5 \times 1.56 = 1.4\,kPa$ (28.5 psf) (long duration);
  $LR1_S = NFL1 \times GTF1 \times LS1 = 1.6 \times 3.8 \times 2.8 = 17.0\,kPa$ (355.8 psf) (short duration).
  $LR2_S = NFL2 \times GTF2 \times LS2 = 1.75 \times 1.0 \times 1.56 = 2.7\,kPa$ (57.1 psf) (short duration)

— The lowest load resistance value controls: $LR = LR2_L$ (long duration) = 1.4 kPa (28.5 psf).

• Determine the applied loads:

— SW load = 18 mm/1000 × 250 kg/m³ × 0.0981 kN/kg = 0.44 kPa (9.17 psf); SL = 1.0 kPa (21 psf); total load = SL + SW = 1.0 + 0.44 = 1.44 kPa (39.2 psf).

- Compare the applied loads to the load resistance (LR):

— 1.44 kPa ≤ 1.4 kPa. The calculated load due to self-weight and snow load is greater than the load resistance by approximately 3%. The designer should decide that the proposed IG panel is only acceptable if a probability of breakage higher than 8 in 1000 is acceptable. The designer may elect to use thicker glass to increase the load resistance and decrease the probability of breakage within acceptable limits.

- Determine the load share on each lite:

— For short duration and long duration loading (LS factors are the same in both cases): Lite 1 carries $[1.56/(2.8 + 1.56)] = 36\%$ of the load and Lite 2 carries $[2.8/(2.8 + 1.56)] = 64\%$ of the load.

- Determine the approximate center-of-glass deflection for each lite, using the lower chart of ASTM E1300 Annex 1, Figures A1.6 and A.1.28 and load shares determined above. The aspect ratio (AR) $= 2438/914 = 2.7$:

— For lite 1: load × area$^2 = 0.36 \times 1.44 \times (0.914 \times 2.438)^2 = 2.4\,\text{kN m}^2$ (5.9 kip ft^2). From the lower chart of Fig. 4.23, $\delta_1 = 5.5$ mm (0.22 in). For lite 2: load × area$^2 = 0.64 \times 1.44 \times (0.914 \times 2.438)^2 = 4.6\,\text{kN m}^2$ (11.1 kip ft^2).

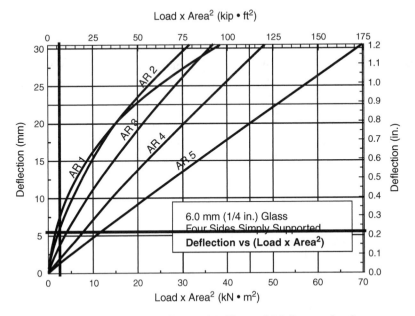

4.23 ASTM E1300, Annex A1, Figure A1.6 (lower chart).

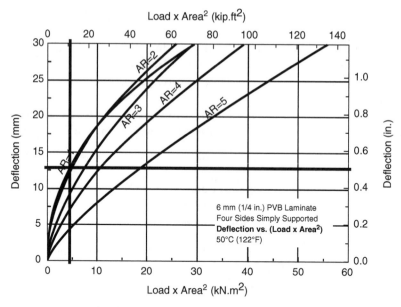

Load x Area² (kip.ft²)

6 mm (1/4 in.) PVB Laminate
Four Sides Simply Supported
Deflection vs. (Load x Area²)
50°C (122°F)

Load x Area² (kN.m²)

4.24 ASTM E1300, Annex A1, Figure A1.28 (lower chart).

4.5.8 Example with four-side supported laminated panel

Determine the load resistance associated with a laminated glass panel described below. Determine if the probability of breakage for a laminated glass panel is less than 1 in 1000 under self-weight of the glass (SW), a given snow load (SL) of 1.44 kPa (30 psf) and a given wind load (WL) of 0.96 kPa (20 psf). In-service laminated glass temperatures do not exceed 50 °C (122 °F). Assume the following:

— Proposed assembly including a nominal thickness 12 mm (1/2 in) annealed (AN) laminated panel. There are screens installed below to prevent falling broken glass.
— Glass dimensions: 1524 mm × 2134 mm (48 in × 72 in).

● Determine the load resistance:

— Determine NFL (nonfactored load) from the upper chart of ASTM E1300 Annex 1, Figure A1.33 (Fig. 4.25); NFL = 6.5 kPa (136 psf).
— Determine the GTF (glass type factor) from ASTM E1300, Table 1 = 0.5 for long duration loads (SW, SL), GTF = 1 for short duration loads (WL) (Table 4.1).
— Determine the load resistance:
— LR_L = NFL × GTF = 6.5 × 0.5 = 3.25 kPa (68 psf) (long duration);

Plate Length (in)

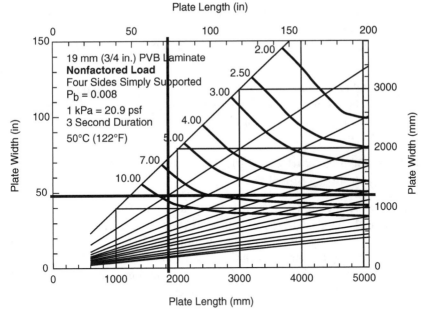

4.25 ASTM E1300, Annex A1, Figure A1.33 (upper chart).

— LR_S = 6.5 ×1 = 6.5 kPa (136 psf) (short duration).
— The long-term load resistance controls: LR_L = 3.25 kPa (68 psf).

• Determine the applied loads:

— SW load = 12 mm/1000 × 2500 kg/m³ × 0.00981 kN/kg = 0.29 kPa (6.1 psf); SL = 1.44 kPa (30 psf); WL = 0.96 kPa (20 psf); total load = SW + SL + WL = 0.29 + 1.44 + 0.96 = 2.69 kPa (58 psf).

• Compare the applied loads to the load resistance (LR): 2.69 kPa ≤ 3.25 kPa. The calculated load due to self-weight and snow load is less than the load resistance. Therefore, the proposed panel is acceptable for the applied loads.
• Check the probability of breakage is less than 1 in 1000:

— Determine the nondimensional lateral load (q^*), using Equation X1.3 in Appendix X1; $q^* = qA^2/Et^4$; t_{actual} = 11.91 mm (0.469 in) from Table 4 (Table 4.2); 2.69 kPa × (3.25 m²)²/[(71.7 × 10⁶ kPa) × (0.01191 m)⁴] = 19.7.
— Determine the probability of breakage using Appendix X3; $P_b = k(ab)^{(1-m)} (Et^2)^m e^J$ (Equation X3.1); t = 11.91 mm; e = 2.7812;

$E = 71.7 \times 10^9$ Pa; $k = 2.86 \times 10^{-53} \text{m}^{12} \text{N}^7$; $m = 7$; $a = 1524$ mm; $b = 2134$ mm.

— Determine J using Figure X3.1 (Fig. 4.26) AR $= 2134$ mm$/1524$ mm $= 1.4$; $J = 9.5$.

The probability of breakage $P_b = k(ab)^{(1-m)} (Et^2)^m e^J = 0.0036$, or 4 in 1000. For the applied loads, the panel must be thicker to achieve a probability of breakage less than 1 in 1000.

4.5.9 Example with two-side supported (parallel edges) monolithic panel

Determine the factored load resistance of a monolithic glass panel under self-weight of the glass (SW), a wind load (WL) of 1.0 kPa (21 psf) and a snow load (SL) of 1.2 kPa (25 psf). In addition, compare results of separate hand-calculated glass stress, including load duration factors, to allowable edge stress shown in Table X9.1 (Table 4.7). Consider load durations of 3 seconds for wind load, 1 month for snow load, and greater than 1 year for the self-weight load. Assume the following:

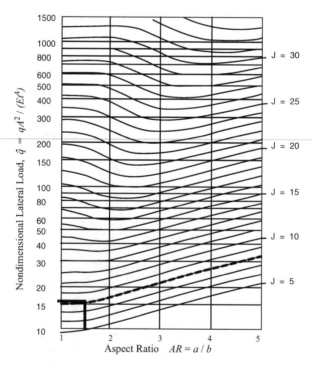

4.26 ASTM E1300, Appendix X3, Figure X3.1.

Table 4.7 ASTM E1300, Appendix X9, Table X9.1

	Clean Cut Edges, MPa (psi)	Seamed Edges, MPa (psi)	Polished Edges, MPa (psi)
Annealed	16.6 (2400)	18.3 (2650)	20.0 (2900)
Heat-strengthened	N/A [A]	36.5 (5300)	36.5 (5300)
Tempered	N/A	73.0 (10 600)	73.0 (10 600)

[A] N/A–Not Applicable.

— Proposed nominal thickness 19 mm (3/4 in) thick annealed (AN) laminated glass with polished edges. There are screens installed below to prevent falling broken glass.
— Glass dimensions with continuous supports along two parallel edges: 914 mm (36 in) (supported edges) × 914 mm (36 in) (unsupported edges).

• Determine and compare the load resistance and applied loads:

— Determine NFL (nonfactored load) from the upper chart of ASTM E1300 Annex 1, Figure A1.41 (Fig 4.27): NFL = 6.94 kPa (145 psf).

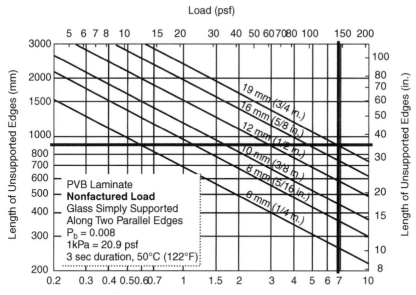

4.27 ASTM E1300, Annex A1, Figure A1.41 (upper chart).

Table 4.8   ASTM E1300, Appendix X6, Table X6.1

NOTE—Calculated to 8/1000 lites probability of breakage (see 3.2.9).

Duration	Factor
3 s	1.00
10 s	0.93
60 s	0.83
10 min	0.72
60 min	0.64
12 h	0.55
24 h	0.53
1 week	0.47
1 month (30 days)	0.43
1 year	0.36
beyond 1 year	0.31

— Determine the GTF (glass type factor) from ASTM E1300, Table 1 (Table 4.1); GTF = 1 for short duration loads and GTF = 0.43 long duration loads.
— Determine LR (load resistance):
   LR_S = NFL × GTF = 6.94 kPa (145 psf) (short duration),
   LR_L = NFL × GTF = 2.98 kPa (62.3 psf) (long duration), controls.
— Determine the applied loads: SW load = $19 \, mm/1000 \times 2500 \, kg/m^3$ ×0.00981 kN/kg = 0.47 kPa (9.8 psf); SL = 1.20 kPa (25 psf); WL = 1.0 kPa (21 psf); total load = SW + SL + WL = 0.47 + 1.2 + 1.0 = 2.67 kPa (55.8 psf).
— Compare the applied loads to the load resistance (LR): 2.67 kPa ≤ 2.98 kPa. The calculated load due to self weight, snow, and wind is less than the load resistance.

• Calculate factored glass stresses and compare with Table X9.1 (Table 4.7).

— Determine load duration factors (LDF) using Table X6.1 (Table 4.8). LDF for beyond a 1 year load duration is 0.31; LDF for a 1 month load duration is 0.43; LDF for beyond a 3 second load duration is 1.0.
— Determine the combined effect of loads of different duration using Equation X7.1 and the LDFs from Table X6.1. The equivalent 3 second load q_3 = SW/0.31 + SL/0.43 + WL/1.0 = 1.52 + 2.79 + 0.29 = 4.6 kPa (96 psf).
— Calculate the maximum flexural moment on the panel: $M = wL^2/8$; w = 4.6 kPa × 0.914 m = 4.20 kN/m; L = 0.914 m; M = 4.20 × $0.914^2/8$ = 0.44 kN m (324 lbf ft).
— Calculate the associated section modulus and flexural stress; note that for long duration loading, conservatively consider the panel as a layered (noncomposite) assembly (see Section 6 for more information on non-, semi-, and full-composite laminated glass unit flexural behavior); t_{actual} per lite = 9.02 mm (0.355 in) by Table 4 (Table 4.2);

$S_x = 2(\text{width} \times t_{\text{actual}}^2/6) = 2(914 \times 9.02^2/6) = 2(12\,394) = 24\,788\,\text{mm}^3$
$(1.65\,\text{in}^3)$; $\sigma = 0.44 \times 1000^2/24\,788 = 17.7\,\text{N/mm}^2(2567\,\text{psf})$.

— Compare calculated maximum stress to allowable edge stress for polished AN glass in Table X9.1 (Table 4.7). $17.7\,\text{N/mm}^2 < 20\,\text{N/mm}^2$; the proposed glass is acceptable.

4.5.10 Example with four-side supported laminated panel

Determine the factored applied loads and compare to the approximate maximum surface stresses associated with a laminated glass panel described below. Determine for a probability of breakage less than 1 in 1000 under self-weight of the glass (SW), a given snow load (SL) of 1.44 kPa (30 psf) and a given wind load (WL) of 1.0 kPa (21 psf). Assume the following:

— The proposed assembly including a nominal thickness 22 mm (7/8 in) annealed (AN) laminated panel.
— Glass dimensions: 1829 mm × 1829 mm (72 in × 72 in).
— In-service laminated glass temperatures do not exceed 50 °C (122 °F).

• Using Table X6.1 (Table 4.8), determine the equivalent short duration (3 second) load value for combined loads of different durations:

— Applied loads: SW load = $22\,\text{mm}/1000 \times 2500\,\text{kg/m}^3 \times 0.00981\,\text{kN/kg}$ = 0.54 kPa (11.3 psf); SL = 1.44 kPa (30 psf); WL = 1.0 kPa (21 psf); total load = SW + SL + WL = 0.54 + 1.44 + 1.0 = 2.98 kPa (62.3 psf).
— Equivalent short duration loads using Equation X7.1 and the LDFs from Table X6.1 (Table 4.8); q_3 = SW/0.31 + SL/0.43 + WL/1.0 = 1.74 + 3.34 + 1.0 = 6.08 kPa (127.3 psf).

• Determine and compare the ratio of applied (q_3) load to nonfactored load (NFL) to determine stresses based on values listed under Equation X8.2 (note that Equation X8.5 shows that Equaion X8.2 stress values are conservative (i.e. higher) with respect to E1300 Section 6 procedures):

— NFL (nonfactored load) from the upper chart of ASTM E1300, Figure A1.12 = 10 kPa (210 psf) (Fig. 4.28).

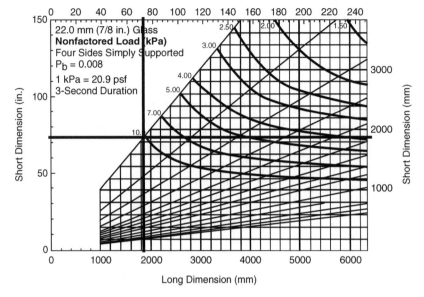

Long Dimension (in.)

4.28 ASTM E1300, Annex A1, Figure A1.12 (upper chart).

— Ratio of applied (q_3) load to NFL $= 6.08/10 = 0.6$.
— By Equation X8.2, a conservative allowable AN surface stress for a 3 second load and a probability of breakage of 8 in 1000 $= 23.3$ MPa (3380 psi). Applied stress, per ratio of applied (q_3) load to NFL $= 0.6 \times 23.3 = 14$ MPa (2028 psi).

• From Equation X8.1, determine the allowable surface stresses associated with a short duration (3 second) load for a probability of breakage of 8 in 1000 and for a probability of breakage of 1 in 1000: $\sigma = (P_b/[k(d/3)^{7/n} \times A])^{1/7}$; $k = 2.86 \times 10^{-53}$ m^{12} N^7, $n = 16$; $A = 1.83^2 = 3.35$ m^2; $d = 3$ s; $P_b = 0.008, 0.001$.

— $\sigma_{8/1000} = (0.008/[2.86 \times 10^{-53} \text{ m}^{12} \text{ N}^7(3/3)^{7/16} \times 3.35])^{1/7} = 13.5$ MPa (1964 psi).
— $\sigma_{1/1000} = (0.001/[2.86 \times 10^{-53} \text{ m}^{12} \text{ N}^7(3/3)^{7/16} \times 3.35])^{1/7} = 10.1$ MPa (1459 psi).
— Compare the applied stress to $\sigma_{8/1000}$ and $\sigma_{1/1000}$; 14 MPa > 13.5 MPa > 10.1 MPa. The applied stress is greater than the allowable stresses associated with either 8 in 1000 or 1 in 1000, so the panel must be thicker to meet the strength requirements.

4.6   Discussion

Glass design for snow loads requires special considerations due to the effects of surface weathering on glass strength, the long duration of applied loads (e.g. snow), and a wide range of in-service temperatures. The following section discusses glass strength, time, and temperature effects to consider when designing glass supporting snow loads.

4.6.1 Glass strength

Glass material strength is dramatically affected by changing surface characteristics. The effect of random surface flaws, nominal surface tensile stresses, contact damage, and water vapor attack can lead to crack propagation (commonly referred to as 'static fatigue' or 'stress corrosion') and tensile stress concentrations initiating fracture under loads. Glass breakage can occur without forewarning, prompting the need to correlate the risk of breakage with a material stress for design. A number of studies describe these characteristics and their probabilistic relation to glass breaking strength; Beason et al. (1998) describe the basis for glass strength utilized in ASTM E1300. Although standard practice bases glass breakage on a probability of 8 in 1000, ASTM E1300 Appendix X3 also provides an optional procedure to determine alternate breakage probabilities. Appendices X8 and X9 list allowable glass stresses associated with a breakage probability of 8 in 1000 under short duration loading. Whereas industry typically considers 8 in 1000 as the basis of design acceptability for vertical glass, some guidelines (e.g. AAMA, 1987) recommend and some codes (e.g. IBC, 2006) require overhead glass design using a probability of 1 in 1000 . As demonstrated in the preceding examples, ASTM E1300 allows designers to provide annealed (AN) glass under long duration loads and breakage probabilities other than 8 in 1000. Study into the effects of static fatigue on heat- and chemically-treated (HS and FT) glass is missing from current research and knowledge, partly due to a wide range of pre-stress values and distributions within standard products allowed by standards.

4.6.2 Load duration (time) effects

Sloped and horizontal glass assemblies, by their orientation to gravity, are more likely to support loads normal to their surface over longer time periods than vertical glass applications. ASTM E1300 defines 'short duration loads' as those that last three seconds or less, whereas 'long duration loads' last 30 days or more. For example, E1300 considers peak (gust) wind loads as short duration, whereas snow and self-weight loads are long duration. ASTM E1300 Appendix, Table X6.1 lists load duration factors (LDFs) for annealed (AN) glass applied to determine glass load resistance due to the effects with other time periods (forthcoming E1300 versions may apply similar LDFs for AN, HS and FT glass kinds). For low slope or horizontally oriented glass, self-weight and snow (and ice) loads are long duration loads and their effects must be considered in combination with short duration loads.

E1300 Appendix X7 provides an approximate technique to determine the combined effect of loads of various durations through load duration factors (LDF). Appendices X8 and X9 list allowable glass stresses associated with a 3 second duration. Load combinations appropriate for AN, HS, and FT glasses are key to providing accurate thickness selection for each glass kind. Load duration factors in Appendix X7 are based on AN glass and are conservative for HS and FT glasses. Some glass engineers design applied tensile stresses to values below common pre-compression stresses in HS and FT glasses to qualify designs (assuming that tensile stresses will not develop), without considering the potential effects of static fatigue over the design life of a glass panel. Research is required to determine if such conditions may arise and if E1300 Table X6.1 load duration factors are appropriate for HS and FT glass.

To account for load duration effects on heat-treated glasses, one may derive LDF values for HS and FT glasses by comparing E1300, Table 1 glass type factor values for AN, HS, and FT glass for short and long duration loads, as follows:

- Determine the ratios of long duration to short duration glass type factors for AN, HS, and FT glass kinds, using E1300, Table 1 (Table 4.1), as follows: AN: 0.5/1.0 = 0.5; HS: 1.3 / 2.0 = 0.65; FT: 3.0 /4.0 = 0.75.
- Derive proposed n-values from E1300, Equation X7.1:
  $q_3 = \text{SUM}\ (q_i[d_i/3]^{1/n})$. Load duration factor $(\text{LDF}) = [d^i/3]^{(1/n)}$; $d^i/3 = (1/\text{LDF})^{(1/n)}$; $n = \log(1/\text{LDF})(d^i/3)/\log(1/\text{LDF})$; proposed n values for AN, HS, and FT glass kinds as follows: AN: 16 (per E1300); HS: 31.7; FT: 47.5.

Based on the above, LDFs for AN from Table X6.1 (Table 4.8) and proposed LDF values for HS and FT glass are as tabulated in Table 4.9.

Temperature effects

Glass in sloped and horizontal applications typically includes laminated assemblies where lamination materials ('interlayers') fuse two or more sheets of glass into a single panel assembly. Laminated panels benefit post-glass breakage safety where interlayer materials may retain fractured glass within their supports and prevent falling glass hazards. Interlayers consist of a

Table 4.9 Proposed load duration factor values for HS and FT glass

Load duration factors					
Time	Seconds	Condition	AN (per ASTM E1300 Table X6.1)	HS (proposed)	FT (proposed)
3 s	3	Wind load	1.00	1.00	1.00
10 s	10		0.93	0.96	0.97
1 min	60		0.83	0.91	0.94
10 min	600		0.72	0.85	0.89
60 min	3600		0.64	0.80	0.86
12 h	43200		0.55	0.74	0.82
24 h	86400		0.53	0.72	0.81
1 week	604800		0.47	0.68	0.77
1 month	2592000	Snow load	0.43	0.65	0.75
1 year	30758400		0.36	0.60	0.71
Beyond 1 year	615168000	Self weight	0.31	0.55	0.67

range of either prefabricated sheets (e.g. ethylene vinyl acetate, polyvinyl butyral (PVB) – the most commonly used interlayer material – or proprietary materials) or cured, liquid-applied materials (e.g. polyester and urethane). The load resistance of a laminated glass assembly depends upon interlayer material stiffness, which in turn depends on its adhesive and cohesive strength. Laminated assemblies develop limited shear transfer from glass to glass across the interlayer due to direct flexure and membrane behavior under loads. Industry efforts are underway to create standard relationships between available interlayer products. Whereas manufacturer's information is available for some interlayers (e.g. PVB) to evaluate composite laminated panel stiffness, information for other products is as yet unavailable and composite stiffness may vary from full- to semi- to noncomposite laminated panel stiffness depending on load duration and material temperature. The designer must consider the specific interlayer materials as an integral part of the safety and serviceability design of laminated glass panels. Behr *et al.* (1993) address the issue of composite bending action in glass panels laminated with PVB interlayer.

Interlayers are viscoelastic materials and their stiffness depends on temperature changes and load duration. Horizontal and sloped glass assemblies commonly experience temperatures ranging from below freezing to over 82 °C (180 °F) through the seasons. In cold temperatures, laminated assemblies may support heavy snow loads for extended periods. Although temperature effects are important to consider in any laminated glass assembly (vertical, high slope, or low slope), the combination of time and temperature effects on low-slope glazing is especially important for safety and serviceability of occupants below.

Interlayer stiffness

Interlayer stiffness is temperature and time dependent, as studied by Bennison *et al.* (1999) and others. ASTM E1300 standard practice includes the use of polyvinyl butyral (PVB) interlayer material properties. Forthcoming versions of ASTM E1300 will include Appendix X11, providing a method to determine effective thickness of laminated glass for analysis of load resistance and deflection in standard practice. The method is intended for use with standard engineering formulae or finite element methods for calculating glass stress and deflection of laminated glass subject to uniform load. Examples below include Appendix X11 procedures.

4.7   Examples employing methods beyond standard practice

The following examples describe procedures for designing glass under snow loads, referencing Appendix X11 in forthcoming versions of ASTM E1300, and proposed load duration factors in Table 4.9 for HS and FT glass kinds.

4.7.1   Example with two-side supported (parallel edges) laminated panel

Determine the equivalent short duration factored load associated with a laminated glass panel under self-weight of the glass (SW) and a snow load (SL) of 2.0 kPa (42 psf). Calculate and compare glass stresses using the effective laminated glass thickness procedure in Appendix X11 to allowable edge stresses shown in E1300, Table X9.1 (Table 4.7). In addition, determine service deflection. Consider load durations of 1 month for snow load and greater than 1 year for self-weight load. Assume the following:

— Proposed nominal thickness 25 mm (1.0 in) strengthened (HS) with seamed edges and a 1.5 mm (0.06 in) thick PVB interlayer.
— Glass dimensions with continuous supports along two parallel edges: 1000 mm (39.4 in) (supported edges) × 1524 mm (60 in) (unsupported edges).

● Determine the effective thickness of the laminated glass panel using Appendix VII:

— The shear transfer coefficient $\Gamma = 1/[1 + 9.6(El_s h_v / G h_s^2 a^2)]$ (Equation X11.1)
— h_v = interlayer thickness 1.5 mm (0.06 in); h_s = thickness associated with laminated panel thickness or single.ply thickness; h_1 = glass ply 1 minimum thickness = 11.91 mm (0.47 in), from E1300, Table 4 (Table 4.2); h_2 = glass ply 2 minimum thickness = 11.91 mm (0.47 in), from E1300 Table 4 (Table 4.2); E = glass Young's modulus = 70 000 MPa (10 400 kip in²); a = length scale (smallest in-plane dimension) = 1000 mm (39.4 in); G = interlayer shear modulus (per Section X11.4) = 0.05 MPa (7.25 psi) (long duration, 1 month, 50 °C (122 °F)).
— $I_s = h_1 h_{s;2}^2 + h_2 b_{s;1}^2 = 11.91 \times 6.71^2 + 11.91 \times 6.71^2 = 1071 \, \text{mm}^3$ (Equation X11.2).
— $h_{s;1} = h_s h_1 / (h_1 + h_2) = 13.41 \times 11.91 / (11.91 + 11.91) = 6.71 \, \text{mm}$ (Equation X11.3).
— $h_{s;2} = h_s h_2 / (h_1 + h_2) = 13.41 \times 11.91 / (11.91 + 11.91) = 6.71 \, \text{mm}$ (Equation X11.4).

— $h_s = 0.5(h_1 + h_2) + h_v = 0.5(11.91 + 11.91) + 1.5 = 13.41\,\text{mm}$
(Equation X11.5).
— $\Gamma = 1/[1 + 9.6(El_s h_v / Gh_s^2 a^2)] = 1/[1 + 9.6(70\,000 \times 1071 \times 1.5)/$
$(0.05 \times 13.41^2 \times 1000^2)] = 0.0083.$
— Determine the effective thickness for the deflection calculation using:

$h_{ef;w} = (h_1^3 + h_2^3 + 12\Gamma I_s)^{1/3}$ (Equation X11.6)

$= (11.91^3 + 11.91^3 + 12 \times 0.0083 \times 1071)^{1/3}$

$= 15.2\,\text{mm}\ (0.60\,\text{in}).$

— Determine the effective thickness for the stress calculation using:

$h_{1;ef;\sigma} = [h_{ef;w}^3 / (h_1 + 2\Gamma h_{s;2})]^{1/2}$ (Equation X11.7)

$= [15.2^3 / (11.91 + 2 \times 0.0083 \times 6.71)]^{1/2} = 17.0\,\text{mm}\ (0.67\,\text{in})$

$h_{2;ef;\sigma} = [h_{ef;w}^3 / (h_2 + 2\Gamma h_{s;1})]^{1/2}$ (Equation X11.8)

$= [15.2^3 / (11.91 + 2 \times 0.0083 \times 6.71)]^{1/2} = 17.0\,\text{mm}\ (0.67\,\text{in}).$

• Determine the applied equivalent short duration load using proposed load duration factors for HS glass:

— Determine the applied loads: SW load $= 25\,\text{mm}/1000 \times 2500$ $\text{kg/m}^3 \times 0.00981\,\text{kN/kg} = 0.62\,\text{kPa}$ (13 psf); SL $= 2.0\,\text{kPa}$ (42 psf).
— Determine load duration factors (LDF) using Table 4.9. LDF for beyond 1 year load duration is 0.55; LDF for a 1 month load duration is 0.65.
— Determine the combined effect of loads of different duration using Equation X7.1. The equivalent 3 second load $q_3 = \text{SW}/0.55$ $+ \text{SL}/0.65 = 1.13 + 2.79 = 3.92\,\text{kPa}$ (81.9 psf).

• Calculate factored glass stresses and compare with Table X9.1 (Table 4.7):

— Calculate the maximum flexural moment on the panel:
$M = wL^2/8$; $w = 3.92\,\text{kPa} \times 1.0\,\text{m} = 3.92\,\text{kN/m}$ (269 plf); $L = 1.524$ m (5 ft); $M = 3.92 \times 1.524^2/8 = 1.14\,\text{kN m}$ (1793 lbf ft).
— Calculate the associated section modulus and flexural stress (note the long duration loading and effective thickness for stress); $s_x = \text{width}$ $\times t_{eff}^2/6 = 1000 \times 17.0^2/6 = 48\,167\,\text{mm}^3\,(2.94\,\text{in}^3)$; $\sigma = 1.14 \times 1000^2/$ $48\,167 = 23.7\,\text{N/mm}^2\,(3437\,\text{psi}).$
— Compare the calculated maximum stress to allowable edge stress in Table X9.1; 23.7 N/mm^2 < 36.5 N/mm^2 (Table 4.7). The proposed

panel is acceptable for the applied load.

- Determine service deflection:

— Calculate the moment of inertia using the effective thickness determined above;
  $I_x = \text{width} \times t_{\text{eff}}^3/12 = 1000 \times 15.2^2/12 = 292\,651\,\text{mm}^4$ $(0.70\,\text{in}^4)$.
— Calculate deflection; $\delta = 5wL^4/(384EI_x)$; $w = 3.92\,\text{kN/m}$ (269 lbf/ft);
  $L = 1.524\,\text{m}$ (5 ft); $E = 70\,000\,\text{MPa}$ $(10\,400\,\text{kip in}^2)$;
  $I_x = 292\,651\,\text{mm}^4$ $(0.72\,\text{in}^4)$; $\delta = 5 \times 3.92 \times 1.524^4/(384 \times 70\,000 \times 1000 \times 292\,651/1000^3) = 13.4\,\text{mm}$ (0.53 in). This is equivalent to a span to deflection ratio of $L/114$.

4.7.2 Example with four-side supported laminated panel

Determine the equivalent short duration factored load associated with a laminated glass panel under self-weight of the glass (SW) and a snow load (SL) of 2.0 kPa (42 psf). Calculate and compare applied and ASTM E1300 Appendix X8 allowable glass stresses using the effective laminated glass thickness procedure in Appendix X11. Consider load durations of 1 month for snow load and greater than 1 year for self-weight load. Assume the following:

— Proposed assembly including a nominal thickness 22 mm (0.87 in) thick fully tempered (FT) laminated panel, consisting of two 10 mm (0.39 in) nominal thickness lites and 1.5 mm (0.06 in) PVB interlayer.
— Glass dimensions: 1219 mm × 1930 mm (48 in × 76 in).
— Interlayer temperatures do not exceed 50 °C (122 °F).

- Determine the effective thickness of the laminated glass panel:

— The shear transfer coefficient $\Gamma = 1/[1 + 9.6(EI_sh_v/Gh_s^2a^2)]$
  (Equation X11.1).
— $I_s = h_1h_{s;2}^2 + h_2h_{s;1}^2 = 9.02 \times 6.75^2 + 9.02 \times 6.75^2 = 499\,\text{mm}^3$
  (Equation X11.2).
— $h_{s;1} = h_sh_1/(h_1 + h_2) = 13.5 \times 9.02/(9.02 + 9.02) = 5.26\,\text{mm}$
  (Equation X11.3).
— $h_{s;2} = h_sh_2/(h_1 + h_2) = 10.52 \times 9.02/(9.02 + 9.02) = 5.26\,\text{mm}$
  (Equation X11.4).
— $h_s = 0.5(h_1 + h_2) + h_v = 0.5(9.02 + 9.02) + 1.5 = 10.52\,\text{mm}$
  (Equation X11.5).
— h_v = interlayer thickness 1.5 mm (0.06 in); h_1 = glass ply 1 minimum thickness = 9.02 mm (0.36 in) from E1300 Table 4 (Table 4.2); h_2 =

glass ply 2 minimum thickness = 9.02 mm (0.36 in) from E1300 Table 4 (Table 4.2); E = glass Young's modulus = 70 000 MPa (10 400 kip in^2); a = length scale (smallest in-plane dimension) = 1219 mm (48 in); G = interlayer shear modulus (per Section X11.4) = 0.05 MPa (7.25 psi) (long duration, 1 month, 50 °C (122 °F)).

— $\Gamma = 1/[1 + 9.6(El_s h_v / Gh_s^2 a^2)] = 1/[1 + 9.6(70\,000 \times 499 \times 1.5)/(0.05 \times 10.52^2 \times 1219^2)] = 0.0161.$

— Determine effective thickness for deflection calculation:
$h_{ef;w} = (h_1^3 + h_2^3 + 12\Gamma I_s)^{1/3}$ (Equation X11.6)
$= [9.02^3 + 9.02^3 + 12 \times 0.0161 \times 499)]^{1/3} = 11.6\,\text{mm}\ (0.46\,\text{in}).$

— Determine effective thickness for stress calculation:
$h_{1;ef;\sigma} = [h_{ef;w}^3/(h_1 + 2\Gamma h_{s;2})]^{1/2}$ (Equation X11.7)

$= [11.6^3/(9.02 + 2 \times 0.0161 \times 5.26)]^{1/2} = 13.0\,\text{mm}\ (0.51\,\text{in})$

● Determine the applied equivalent short duration load using proposed load duration factors for HS glass:

— Determine the applied loads: SW load = 22 mm/1000 × 2500 kg/m^3 × 0.00981 kN/kg = 0.54 kPa   (11.3 psf); SL = 2.0 kPa (42 psf).

— Determine load duration factors (LDF) using Table 4.9. LDF for beyond a 1 year load duration is 0.55; LDF for a 1 month load duration is 0.65.

— Determine the combined effect of loads of different duration using Equation X7.1. The equivalent 3 second load $q_3 = $ SW/0.55 +SL/0.65 = 0.98 + 3.08 = 4.06 kPa (85 psf).

● Check if the probability of breakage is less than 1 in 1000:

— Determine the nondimensional lateral load (q^*), using Equation X1.3 in Appendix X1: $A = 2.97 \times 10^6$ mm; $E = 70\,000$ MPa (10 400 kip in^2); $t = h_{1;ef;\sigma} = 13.0$ mm (0.51 in); $q^* = qA^2/Et^4 = 4.06 \times (2.35 \times 10^6)^2/(70 \times 10^6 \times 13.0^4) = 11.1.$

— Determine the probability of breakage using Appendix X3:
$P_b = k(ab)^{(1-m)}(Et^2)^m e^j$ (Equation X3.1); $t = 13.0$ mm; $e = 2.7812$; $E = 70 \times 10^9$ Pa; $k = 2.86 \times 10^{-53}$m^{12}N^7; $m = 7$; $a = 1219$ mm; $b = 1930$ mm.

— Determine J using Figure X3.1 (Fig. 4.29): AR = 1930 mm/1219 mm = 1.6; $J = 5.1.$

— The probability of breakage $P_b = k(ab)^{(1-m)}(Et^2)^m e^j = 0.00094$, or 0.9 in 1000. For the applied loads, the panel is sufficient for a probability of breakage less than 1 in 1000.

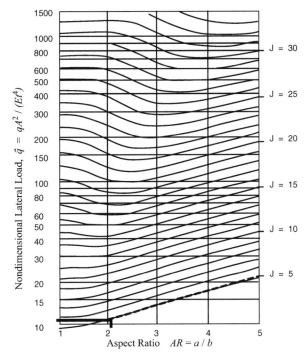

Nondimensional Lateral Load, $\hat{q} = qA^2/(Et^4)$

Aspect Ratio $AR = a/b$

4.29 ASTM E1300, Appendix X3, Figure X3.1.

- From Equation X8.1, determine the allowable surface stresses for the proposed panel associated with a short duration (3 second) load, for probabilities of breakage of 8 in 1000 and 1 in 1000. $\sigma_{AL} = (P_b/[k(d/3)^{7/n} \times A])^{1/7}$; $k = 2.86 \times 10^{-53}$ m^{12}N^7, $n = 16$; $A = 1.22 \times 1.93 = 2.35$ m^2; $d = 3$ seconds; $P_b = 0.008, 0.001$.

— $\sigma_{AL,8/1000} = (0.008/[2.86 \times 10^{-53}\text{m}^{12}\text{N}^7(3/3)^{7/16} \times 2.35])^{1/7}$
   $= 14.2$ MPa (2260 psi).
— $\sigma_{AL,1/1000} = (0.001/[2.86 \times 10^{-53}\text{m}^{12}\text{N}^7(3/3)^{7/16} \times 2.35])^{1/7}$
   $= 10.6$ MPa (1537 psi).

4.8   Conclusions

This chapter describes glass design to resist snow loads, including safety and serviceability design. Included is a design strategy to avoid water penetration when snow and ice accumulation may occur. The chapter also includes a description and glass design examples employing United States standard practice to determine glass load resistance under snow loads, and additional considerations for acceptable serviceability performance. The chapter discusses static fatigue effects on glass strength, load duration, and temperature effects on laminated glass assemblies, including examples employing advanced, nonstandard design approaches for their combined effect. Cited references provide background and additional in-depth information.

4.9   Acknowledgement

ASTM (American Society of Testing and Materials) (2007) Figures are reprinted, with permission, from E1300 - 07^{e1}, *Standard Practice for Determining Load Resistance of Glass in Buildings*, copyright ASTM International, 100 Barr Harbor Drive, West Conshohocken, PA 19428. A copy of the complete standard may be obtained from ASTM, www.astm. org.

4.10   References

AAMA (American Architectural Manufacturers Association) (1987) *Glass Design for Sloped Glazing*, AAMA, Des Plaines, IL.

ASCE (American Society of Civil Engineers) (2005) *Minimum Design Loads for Buildings and Other Structures*, ASCE/SEI 7-05, American Society of Civil Engineers, Reston, VA.

ASTM (American Society of Testing and Materials (various dates), E283 – 04, *Standard Test Method for Determining Rate of Air Leakage Through Exterior Windows, Curtain Walls, and Doors Under Specified Pressure Differences Across the Specimen*; E330-02, *Standard Test Method for Structural Performance of Exterior Windows, Doors, Skylights and Curtain Walls by Uniform Static Air Pressure Difference*; E331-00, *Standard Test Method for Water Penetration of Exterior Windows, Skylights, Doors, and Curtain Walls by Uniform Static Air Pressure Difference*, ASTM International, West Conshohocken, PA.

Beason, L., Kohutek, T. and Bracci, J. (1998) Basis for ASTM E1300 Annealed Glass Thickness Selection Charts, *Journal of Structural Engineering*, 124(2), 215–221.

Behr, R. A., Minor, J. E. and Norville, H. S. (1993) The structural behavior of architectural laminated glass, *ASCE Journal of Structural Engineering*, 119 (1), 203–222.

Bennison, S. J., Jagota, A. and Smith, C. A. (1999) Fracture of glass/polyvinyl butyral (Butacite®) laminates in biaxial flexure, *Journal of the American Ceramic Society*, 82(7), 1761–1770.

ICC (International Code Council) (2006) *International Building Code*, Delmar Cengage Learning, Country Club Hills, IL.

Vigener, N. and Brown, M. (2006) Building envelope design guide – sloped glazing, in *The Whole Building Design Guide*. Available at www.wbdg.org/design.

5

Wind pressures on building envelopes

K . C . M E H T A , Texas Tech University, USA

Abstract: Building envelopes of glass are subject to wind-induced loads. Wind pressures for design purposes are obtained following applicable building codes or standards. Details of wind pressures given in the ASCE 7-05 standard and commentary on parameters of these pressure criteria are given in this chapter. An example is also included to illustrate determination of wind loads on components of the building envelope.

Key words: wind pressures, ASCE 7-05 standard criteria, wind loads on components, example of wind loads.

5.1   Introduction

Building envelopes, whether they are made of glass, aluminum panels, marble slabs, or another material, are subject to environmental loads including wind effects. In windstorms, such as hurricanes, tornadoes, and thunderstorms it is common to experience damage to the building envelope. Damage to building envelopes that occurred in Houston, Texas, as a result of Hurricane Alicia in 1983 is shown in an aerial view in Fig. 5.1. This kind of damage to the building envelope culminates in additional damage to the interior and contents of the building, resulting in large property loss and disruption of normal building functions. This chapter focuses on wind pressures that are generated on building envelope components in high winds. Window glazing systems to resist these pressures are given in subsequent chapters.

Wind pressures for design purposes are obtained following applicable building codes or standards. The most prevalent building code used in the United States is the *International Building Code* (ICC, 2006). This building code is adopted by most jurisdictions in the country with perhaps some changes in some jurisdictions. For wind pressures, though, changes by jurisdictions are minor. The *International Building Code* adopts the wind loads specified in the American Society of Civil Engineers Standard ASCE 7

5.1 Damage to glass windows in Houston, Texas, resulting from Hurricane Alicia in 1983 (courtesy of WISE, Texas Tech University).

(ASCE, 2005). This adoption makes wind loads consistent across the country. There is another building code developed by National Fire Protection Agency in recent years, NFPA 5000 (NFPA, 2006), though its adoption by local jurisdictions is not widespread. Fortunately, NFPA 5000 also uses wind loads given in ASCE 7, thus making wind pressure criteria across the country uniform.

The above paragraph suggests that a good knowledge of wind pressures specified in ASCE 7 allows design professionals to determine wind loads for professional practice. The latest version of the ASCE 7 was published in 2005, simply referred to as ASCE 7-05 in this book. Details of wind pressures given in this standard and commentary on parameters of these pressure criteria are the subject of this chapter. An example is also included in the chapter to illustrate determination of wind loads on components of the building envelope.

5.2   Evolution of ASCE 7-05

A brief historical background of the ASCE 7 standard will help to set the stage for more details of wind pressures. The ASCE 7 publication, SEI/ASCE standard, *Minimum Design Loads for Buildings and Other Structures,* is a consensus standard. It originated in 1972 when the American National Standards Institute published a standard ANSI A58.1 with the same title (ANSI, 1972). That 1972 standard was revised 10 years later, containing an innovative approach to wind loads for components and cladding (C&C) of buildings (ANSI, 1982). Wind load criteria in the ANSI A58.1-1982 were based on understanding of aerodynamics of wind pressures in building corners, eaves, and ridge areas, as well as the effects on pressures from averaging of areas (size of a component).

In the mid-1980s, ASCE assumed responsibility for the design load standard. The document published as ASCE 7-88 (ASCE 1990) continued design load criteria for live loads, snow loads, wind loads, earthquake loads, and other environmental loads, as well as load combinations the same as ANSI A58.1-1982. The ASCE Standards Committee has voting membership of close to 100 individuals from many backgrounds, including consulting engineering, architecture, research, construction industry, education, government, design, and private practice. The criteria for each of the environmental loads are developed by respective task committees and are voted by the full standard committee.

The wind load criteria of ASCE 7-88 (ASCE, 1990) were essentially the same as ANSI A58.1-1982. In 1995, ASCE published ASCE 7-95 (ASCE, 1995). This 1995 version contained major changes in wind load criteria; the basic wind speed averaging time was changed from fastest-mile to 3 second gust. This, in turn, necessitated significant changes in boundary layer profile parameters, gust effect factors, and some pressure coefficients. *Guide to the Use of the Wind Load Provision of ASCE 7-95* (Mehta and Marshall, 1997) was published by ASCE to assist practicing professionals in the use of wind load criteria of ASCE 7-95.

In the year 2000, ASCE published a revision of ASCE 7-95 with updated wind load provisions. The document has the same title and was termed ASCE 7-98 (ASCE, 2000). The *International Building Code* (ICC, 2000) adopted wind load criteria of ASCE 7-98 by reference. This was a major milestone since it established a single wind load criterion for all buildings and structures for the entire United States. Subsequently, ASCE revised the standard with some changes in wind load criteria in ASCE 7-02 (ASCE, 2002) and ASCE 7-05 (ASCE, 2005); each of the revisions is adopted by *International Building Code* updates. The wind load criteria used in this chapter are of ASCE 7-05 (i.e. the ones contained in IBC 2006).

5.3   Standard of practice for wind pressures

Wind load is a complex subject because it depends on many factors. A primary goal of a wind load standard is to provide wind loads that are consistent irrespective of the location of the building or the size and shape of the building. However, design wind speed varies with probability of occurrence (mean recurrence interval), risk of failure that is acceptable, terrain surrounding the building, and other unknown factors such as effects of mountains and valleys. In addition, wind loads are affected by shape of the building, location of a component on the building, and area of the component on which load is acting. A wind load standard such as ASCE 7 attempts to take into account all these factors and provide wind loads through simple equations that are easy to use by practicing professionals.

The equations to calculate design wind pressure, p, are

$$p = qGC \quad (\text{psf or N/m}^2) \tag{5.1}$$

and

$$q = 0.00256\, K_z\, K_{zt}\, K_d\, V^2\, I \quad (\text{psf}) \tag{5.2}$$

[in SI: $q = 0.631\, K_z\, K_{zt}\, K_d\, V^2\, I \quad (\text{N/m}^2)$]

where

q = effective velocity pressure (psf or N/m^2)
G = gust effect factor
C = pressure coefficient
K_z = exposure velocity pressure coefficient reflecting height above ground and surrounding terrain
K_{zt} = topographic factor
K_d = directionality factor
I = importance factor
V = basic wind speed from the map (mph or m/s)

The ASCE 7 standard permits use of a wind tunnel when geometry of the building is complex and the design of the frame and/or glazing system is important. Wind pressures specified in the standard for components and cladding are obtained from wind tunnel tests of box shape models and thus have limitations in their use. A wind tunnel is able to model the shape of the building and surrounding buildings or other structures. Simulated winds are used in a wind tunnel that mimics natural wind, but not windstorms per se. Thus, it is important to realize that wind tunnel tests have limitations. Technology of wind tunnel tests continues to improve and the tests continue to provide more accurate wind pressures.

Technology of computer simulation of wind pressures, computational fluid dynamics (CFD) is an evolving field. CFD commercial software packages are available that provide wind pressure values on different shapes. However, accuracy of the results may not be very good in all cases. Professional practitioners need to use extreme caution when using the results of CFD for design of buildings and structures.

5.4   Basic wind speed

Wind speed is the most important parameter in determination of wind pressures because it is the square of the wind speed V that is used in the equation for wind pressure. The equations shown above convert kinetic energy of wind into potential energy in terms of pressure.

Basic wind speed given in the map of the ASCE 7 standard is a 3 second gust speed in open terrain (e.g. airport and open field) at 10 meters (33 ft) above ground. It is also associated with a 50 year mean recurrence interval (MRI). It is important to have some understanding of the wind speed values given in the ASCE 7 standard.

The first item critical about wind speed is the averaging time, such as the 3 second gust speed. There are other averaging times used in the literature and media that affect the numerical value of the wind speed. The National Hurricane Center uses a 1 minute averaging time for hurricane wind speed, often called sustained wind speed. In the ASCE 7 standard prior to 1995 and in the ANSI standard of the 1980s and 1970s, wind speed values were given as the fastest-mile wind speed. The term 'fastest-mile' represents one mile of wind passing by a point in the shortest time. The averaging time for the fastest-mile wind speed varies with the magnitude of the wind speed; this is explained below. In Canada, design wind speeds are specified as the mean hourly wind speed, while some countries in the Asia-Pacific rim use the 10 minute wind speed. In each case the numerical value of wind speed will be different for the same wind speed record.

An old strip chart of a wind speed record obtained by the National Weather Service (NWS) is shown in Fig. 5.2. As can be seen in the figure, the wind speed fluctuates continuously and randomly. The horizontal axis in the figure is time; each division represents 5 minutes. The vertical axis in the figure is the wind speed value in knots (1 knot = 1.15 mph or 0.514 m/s). The peak gust recorded is close to 86 knots (99 mph or 44.2 m/s), while the 5 minute average may be around 65 knots and the hourly wind speed value would be even lower. It is difficult to determine the wind speed value for different averaging times from a visual graph such as Fig. 5.2.

Conversion of wind speed from one averaging time to another can be accomplished by use of a curve given in the Commentary of the ASCE 7 standard, reproduced in Fig. 5.3. The curve gives the averaging time of wind speed on the horizontal axis and the ratio of wind speeds of specific averaging time to mean hourly wind speed (3600 seconds) on the vertical axis. Thus, the ratio of the wind speed for one hour averaging time (mean hourly wind speed) is 1.0. The ratio of the 3 second averaging time is 1.52, while the ratio of the 1 minute (60 seconds) averaging time is 1.25. The fastest-mile wind speed has varying averaging times depending on the value of the wind speed. For example, the 60 mph fastest-mile wind speed will

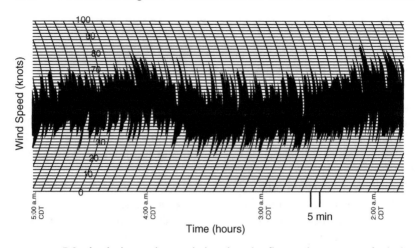

5.2   A wind speed record showing the fluctuating nature of wind.

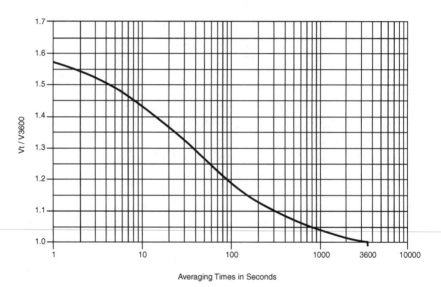

Averaging Times in Seconds

5.3   Wind speed versus averaging time (reproduced from ASCE7.05 with permission from American Society of Civil Engineers).

have an averaging time of 60 seconds while the 120 mph fastest-mile wind speed will have an averaging time of 30 seconds. In essence, the averaging time in seconds for the fastest-mile wind speed is the ratio of 3600 divided by the wind speed value in mph. There has never been a fastest-kilometer wind speed measured.

Equivalent wind speed values for a mean hourly wind speed of 70 mph for different averaging times are obtained using Fig. 5.3 and are given below.

- Mean hourly wind speed is 70 mph (31.3 m/s).
- Ten minute wind speed will be $70 \times 1.07 = 75$ mph (33.5 m/s).
- One minute wind speed will be $70 \times 1.25 = 88$ mph (39.3 m/s).
- Three second wind speed will be $70 \times 1.52 = 106$ mph (47.6 m/s)

The fastest-mile wind speed will be 91 mph (which requires iteration by assuming the value of the fastest mile and determining the averaging time prior to going to the curve).

Another significant factor in determining wind speed is the probability of occurrence of the basic design wind speed. Definition of a 50 year mean recurrence interval (MRI) is the probability of occurrence of 0.02 in any one year. Basic wind speed values of the ASCE 7 standard shown in the map are for a 50-year MRI and are obtained from statistical analysis of the recorded historical data. Statistical analysis and its confidence are highly dependent on the number of years of recorded data. Most of the NWS data are archived at the National Climatic Data Center (NCDC) in Ashville, NC, and are of the length of 40 to 60 years for a location. An increase in length of the record to reduce variability (better confidence) in statistical analysis is accomplished by combining data of a number of stations (generally 8 to 10) for a given region. This procedure provides data of 300 to 400 years in length for a region. The wind speed records were checked to make sure that the data of stations in a region were independent prior to combining them. Regional stations, called Super Stations, varied in number from 40 to 50 in the country depending on the size of the regions chosen. Extreme value statistics are used to assess the probability of occurrence of wind speed. In the middle of the country wind speed associated with a 50 year MRI came close to 90 mph with small variations. In the West Coast area the wind speed value for a 50 year MRI came close to 85 mph.

Along the Gulf Coast and Atlantic Coastline hurricane winds control the occurrence of high winds. Since there is not enough hurricane wind data recorded at any given location along the coast, the Monte Carlo simulation procedure is used to establish wind speeds related to the probability of occurrence. Briefly, a Monte Carlo simulation captures hurricane characteristics of the past 100 years of data of hurricanes and predicts hurricane strikes along the coastal areas for thousands of years. This simulation provides hurricane wind speed data for statistical analysis similar to inland NWS recorded wind speed data.

Statistical analysis provides a wind speed for any annual probability of occurrence, e.g. an annual occurrence of 0.01 (100 year MRI) or 0.001 (1000 year MRI). The probability of exceeding a given wind speed during the life of a building is different than the annual occurrence of wind speed. A simple Poisson's distribution correlates the annual probability of exceeding wind speed to the probability of exceeding during a reference continuous period

Table 5.1 Probability of exceeding the design wind speed, P_n, during the life of a building

		Design life of structure n (years)					
P_a	MRI	1	5	10	25	50	100
0.04	25	0.04	0.18	0.34	0.64	0.87	0.98
0.02	50	0.02	0.10	0.18	0.40	0.64	0.87
0.01	100	0.01	0.05	0.10	0.22	0.40	0.64
0.002	500	0.002	0.01	0.02	0.05	0.10	0.18
0.001	1000	0.001	0.005	0.01	0.02	0.05	0.10

(life of a building). The equation given in the commentary of the ASCE 7-05 is

$$P_n = 1 - (1 - P_a)^n \qquad [5.3]$$

where P_n is the probability that the wind speed will be exceeded during a given life (n years) of the building, P_a is the annual probability of exceeding (reciprocal of MRI), and n is the life of a building in years.

Using this equation, a table of probabilities can be assembled (see Table 5.1). A wind speed value associated with a 50 year MRI has a 64% chance of exceeding during the 50 year life of a building, while it has an 87% chance of exceeding if the life of the building is 100 years. If a 100 year MRI wind speed is used in the design it has a 40% chance of exceeding during the 50 year life of the building. Wind speeds exceeding the design value do not necessarily cause failures because of safety factors (uncertainty factors) used in the structural design.

5.5   Effective velocity pressure, q

The equation for effective velocity pressure, q shown in Equation [5.2], adjusts basic wind speed of the map for terrain surrounding the building, height above ground, direction of wind attack, and topography. It also converts wind speed into pressure. The constant 0.00256 includes the standard density of air and dimensional conversion of miles per hour (mph) to feet per second (fps).

The exposure velocity pressure coefficient, K_z, reflects the change in wind speed due to the height above ground and surrounding terrain. Wind speed at a gradient height of 1000 to 1500 ft (300 to 460 m) above ground is not affected by the ground roughness. Friction of wind with ground slows the wind near the ground. The rate of reduction in wind speed depends on the level of friction and height above ground. The rough terrain of a suburban area slows the wind more than the smooth terrain of flat and open country.

Terrains are divided into three categories in ASCE 7-05 depending on the ground roughness: Exposure Category B for suburban and wooded terrain; Exposure Category C for flat, open terrain; and Exposure Category D for water. An empirical power law is used to describe the terrains in mathematical terms. The power law exponent and gradient height for each terrain category are given in ASCE 7-05. Wind speeds given in the map are for Exposure Category C (flat, open terrain). These wind speeds are adjusted for height above ground and exposure categories by use of equations (shown in the Commentary of ASCE 7-05). A table of values for K_z is developed and is given in the standard. ASCE 7-05 gives a description and minimum distance necessary in the upwind direction for terrain Categories B and D; if the upwind terrain does not fit the definitions of Categories B and D, the default terrain Category C has to be used.

The topographic factor, K_{zt}, accounts for the speed-up of wind speed over hills, ridges, and escarpments. When wind travels over hills and escarpments it will speed up near the top of the topographic feature. This speed-up depends on the slope of the topographic feature, horizontal distance from the top, and height above the ground where the building is located; these factors make it very complex to quantify the increase in wind speed. ASCE 7-05 gives a combination of equations and a table to specify K_{zt} values. Fortunately, the topographic effect is to be used only for an isolated hill or escarpment. If there are similar topographic features upwind within 100 times the height of the topographic feature or within 2 miles, the factor K_{zt} will have a value of 1.0. Thus, most often rolling hills topography does not require the use of the topographic factor and the value of $K_{zt} = 1.0$.

The directionality factor, K_d, is used to compensate for a smaller chance of wind coming from a specific direction and causing maximum pressure on a building and its components. In versions prior to the ASCE 7-98 standard the directionality was hidden in the load combination formulation. In ASCE 7-05 there is a table that gives values for K_d. Because load combinations are adjusted by this factor, there is a requirement that this factor should be used only when load combinations of ASCE 7-05 are used. There is a controversy in professional practice in the design of window glass in hurricane prone regions whether the directionality factor should be used or not. It is argued that hurricane winds affect a building from more than one direction during its passage. Also, the design of window glass does not use load combination of two loads (i.e. in the combination of dead load plus wind load, dead load is negligible). These arguments would suggest that the directionality factor, K_d, should not be used for the design of window glass. Since the directionality factor K_d is always less than one, it is conservative to use a value of 1.0. Technically, wind load criteria of ASCE 7-05 have considered the smaller chance of wind coming from a specific direction and an even smaller chance of causing maximum pressure at a given location on the building. There is no requirement or guidance in the ASCE 7-05 standard that suggests ignoring the directionality factor. The strength of window glass has a large coefficient of variation. Also, under specific design loads failure of a few windows is acceptable. This combination of large variability in strength and accepted level of window glass failures has resulted in a mandate of not using the directionality factor by some local building code officials. This controversy will not be solved until professional practice of the design of window glass uses limit state design procedures with specific uncertainty factors (load and resistance factors) in the loading side and in the resistance side.

The importance factor, I, accounts for adjusting wind speed for different probabilities of occurrence of wind or MRIs. In ASCE 7-05, wind speeds associated with different MRIs are tied to the importance of a building in terms of consequences of a failure. Buildings that have a low hazard to human life, such as agricultural buildings or temporary buildings, can use wind speeds related to a 25 year MRI. Buildings that have a high hazard to human life, such as sport venues or other buildings with more than 300 people congregated in an area, are required to design for wind speeds related to a 100 year MRI. As indicated in Section 5.4, wind speed with a 100 year MRI has a smaller chance of exceeding during the life of a building. ASCE 7-05 gives values of importance factors, I, in a table and guidelines for its use for different types of buildings. Specific building uses are given in the standard, though it does not cover all building uses; professional judgment needs to be made for each building. Wind speed associated with a 100 year MRI increases the load by 15% in design; this is equivalent to an increase in wind speed of 7%. Consequences of failure of a building or a component should be considered in the design of each building.

5.6   Design pressures for components and cladding, p

Architectural glazing falls under the definition of components and cladding (C&C) in ASCE 7-05. Glazing is not the main wind force resisting system (MWFRS), or frames that provide overall support and stability for the building, hence it is components and cladding. For C&C, Equation [5.1] is expanded in ASCE 7-05 to

$$p = q(GC_p) - q_i(GC_{pi}) \quad (\text{psf or N/m}^2) \qquad [5.4]$$

where

q_i = effective velocity pressure q_z or q_h (details discussed below)
(GC_p) = combined gust effect factor and external pressure coefficient
(GC_{pi}) = combined gust effect factor and internal pressure coefficient

The effective velocity pressure, q_z, is obtained when K_z is evaluated at a specific height above ground, z, and is applied in Equation [5.2]. The value of q_h is obtained when K_h is evaluated at mean roof height, h, and is applied in Equation [5.2]. Since the values of exposure velocity pressure coefficients are obtained from a table in ASCE 7-05 for different terrains and heights, K_z and K_h are obtained easily.

Pressure coefficients, external or internal, that are given in ASCE 7-05 are obtained from wind tunnel tests. Full-scale measurements in the field are conducted on a very limited basis to validate pressure coefficients obtained in wind tunnel tests (Yeatts and Mehta, 1993). For C&C tests in a wind tunnel, pressure results are normalized to obtain nondimensional coefficients that are a combination of the gust effect factor and the pressure coefficient, (GC). This is the reason for specifying the combined (GC) in ASCE 7-05; they cannot be separated. Whenever combined gust effect factor and pressure coefficients (GC) are given, they should be used as a combined value in Equation [5.4].

In addition, results of wind tunnel tests normalize measured pressures with effective velocity pressures at the roof height of the building, q_h, or at a height above ground, q_z. This makes it mandatory that an appropriate value of q is used in Equation [5.4]. For a windward wall receiving positive external pressures, ASCE 7-05 specifies the use of an effective velocity pressure, q_z, evaluated at height z above ground. The effective velocity pressure, q_h, is specified to be used in Equation [5.4] for leeward and side walls and for roofs that receive negative external pressures. Implications of this use of different effective velocity pressures are that pressures for C&C in a windward wall (positive pressures) vary with height, while the same for leeward wall and side walls (negative pressures) are uniform for the entire building height.

The effective velocity pressure, q_i is used along with (GC_{pi}) in Equation [5.4] for an assessment of the internal pressure. The effective velocity pressure q_i can be q_z or q_h. In most cases, the definition of q_i will be q_h determined at the building roof height. An exception of the use of q_z for q_i is only when there is likely to be specific openings in the building at some height above ground. ASCE 7-05 defines the use of the appropriate q_z or q_h associated with (GC_{pi}). In mid-rise buildings for heights up to 250 ft (75 m) the difference between use of q_h and q_z for internal pressures is likely to be less than 10%. Use of q_h to determine an internal pressure is conservative.

ASCE 7-05 gives combined gust effect factor and external pressure coefficient (GC_p) values in a graphical form in Figure 6-17 of the standard, which is reproduced here in Fig. 5.4. This figure will be used in the illustrative example given in the next section. This figure gives (GC_p) values for buildings with a mean roof height, h, greater than 60 ft. For low-rise buildings with a mean roof height of less than 60 ft, a separate figure is given

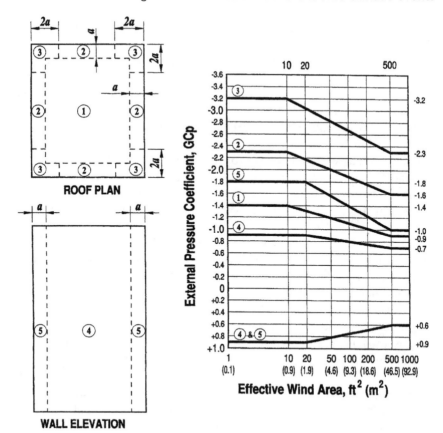

ROOF PLAN

WALL ELEVATION

5.4 Combined gust effect factor and external pressure coefficient for C&C of buildings with a mean roof height $h > 60$ ft (reproduced from ASCE 7-05 with permission from American Society of Civil Engineers).

in ASCE 7-05. In Fig. 5.4, (GC_p) values are related to the effective wind area of the C&C. The effective wind area of a component can be a tributary area of a C&C, though it is defined as the span of the C&C multiplied by width, which need not be less than one-third of the span. When a component is supported on all four sides, such as typical window glass installations, use of its tributary area is appropriate for an effective wind area. When a component is supported at its ends, its wind effective area will be span multiplied by width. The effective wind area for a component is used to obtain wind pressure. The example given in the next section illustrates determination of the effective wind area. The provision that width need not be less than one-third of the span is to account for the manner in which effective wind areas were determined in wind tunnel tests. It should be noted that the effective wind area is to be used only for obtaining the value of (GC_p); the tributary area should always be used in the design of the member.

Values of the external pressure coefficient, (GC_p), vary significantly with the effective wind area, as shown in Fig. 5.4. This variation reflects the fluctuating nature of wind pressures and averaging effects over an area. The larger the effective wind area, the smaller is the value of (GC_p). In addition, values of (GC_p) are different for wall corners and the middle of the wall area. As indicated in Fig. 5.4, the wall corner Zone 5 has larger coefficient values than the middle part of the wall, Zone 4. At wall corners the wind separates from the building corner causing extra turbulence and higher outward acting pressures at the separating corner. Negative pressures (outward acting pressures) are higher for corner Zone 5 as compared to middle Zone 4, while positive pressures acting toward the surface are the same for Zones 4 and 5. In Fig. 5.4, the effective wind area has a log scale, which makes it difficult to read the value of the effective wind area. Equations for these curves can be written for use with a calculator (Mehta and Delahay, 2004) or by employing commercial software, e.g. software published by SDG, Inc. (Morse *et al.*, 2005).

Internal pressures are a result of wind pressure waves acting through openings in the building envelope. All buildings have some openings because of leakage in the envelope as well as openings for air exchange and other breaches in the roof. A result of these openings is fluctuating internal pressure that can act toward the building envelope surface (positive internal pressure) or act away from the surface (negative internal pressure). In determination of the total net pressure on C&C of the envelope ASCE 7-05 requires the use of positive and negative internal pressures, whichever result in the larger pressure. When there is a dominant opening in one of the walls, the magnitude of internal pressure increases significantly. The increase in internal pressure depends on a large (dominant) opening in one wall as compared to small openings in other walls and the roof. However, if the other walls have large openings the internal pressures are not high. The problem of internal pressure is quite complex; ASCE 7-05 simplifies this complex problem by defining buildings as 'enclosed' and 'partially enclosed'. Internal pressures for a partially enclosed building are almost three times higher than that for an enclosed building.

Architectural glazing, where windows are not operable, may be considered as an enclosed building. If the windows are operable and/or if there are sliding doors for balconies, which can come off the guide in high winds, the building may need to be considered as a partially enclosed building. This results in higher internal pressure. Also, if there is a potential of debris that can break window glass, the building is likely to experience high internal pressure during a windstorm.

Internal pressure coefficient, GC_{pi}, values are tabulated in Figure 6-5 of ASCE 7-05. A decision has to be made by the practicing professional whether to consider a building enclosed or partially enclosed. At the outset it should be understood that the building falls in the category of partially enclosed; if not, it is enclosed (default value). A building is considered as partially enclosed when one of the walls has or there is potential of dominant openings as compared to other walls and the roof. The building needs to satisfy both of the following criteria for a partially enclosed building:

1. The total area of openings in a wall that receives positive external pressure exceeds the sum of the areas of openings in the balance of the building envelope by more than 10%.
2. The total area of openings in a wall that receives positive external pressure exceeds 4 ft^2 or 1 % of that wall and the percentage of the openings in the balance of the building envelope do not exceed 20 %.

The above requirements essentially indicate that when there is an opening in one wall that is more than the sum of opening areas in other walls, the air pressure is likely to be trapped inside the building and cause high internal pressure.

Use of Equations [5.2] and [5.4] to determine wind pressures on window glass components and supporting mullions is illustrated in the following example.

5.7   Example for cladding pressures

This example illustrates determination of cladding pressures on a 20 story building using the provisions of ASCE 7-05 (ASCE, 2005). Some discussion is given to ascertain terrain roughness, building category, enclosure classification, etc. However, it is impossible to cover all cases in one example. Since the ASCE 7-05 wind load standard is used throughout this example, it will only be referred to as 'Standard' for brevity purposes.

5.7.1 Example building data

Dimensions: 20 story building; floor height of 11 ft
Footprint of 120 ft × 200 ft with a cutout of 40 ft × 40 ft (see Fig. 5.5)
Height of 220 ft
Flat roof
Framing: Rigid frame in both directions (does not impact determination of cladding pressures)
Cladding: Mullions span 11 ft between floors; spacing of 6 ft
Glazing panels are 6 ft wide × 5 ft 6 in high
Location: Memphis, Tennessee

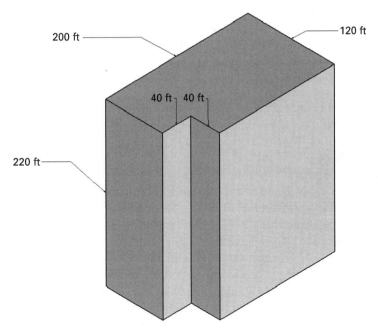

200 ft

120 ft

40 ft   40 ft

220 ft

5.5 Example building dimensions.

Terrain:	Building is located on the Mississippi River bank; it is surrounded by suburban area
Topography:	Homogeneous
Building Category:	The building function is office space. It is not considered an essential facility in case of a disaster and will not have 300 or more people in one area. Building Category II is appropriate for this building, as indicated in Table 1-1 of the Standard. The importance factor = 1.0.
Basic wind speed:	Memphis, Tennessee, is in the interior of the country (not a hurricane-prone region or a special wind region). The basic wind speed is obtained from Figure 6-1 of the Standard. Wind speed V = 90 mph.

5.7.2 Calculating different pressures

Velocity pressures are determined using Equation 6-15 of the Standard:

$$q_z = 0.002\,56\,K_z\,K_{zt}\,K_d\,V^2\,I \qquad [5.2]$$

where V and I are given above; 0.002 56 is a constant; other terms are discussed below.

K_z is the velocity pressure exposure coefficient, which depends on the surrounding terrain. Since the building is located on the Mississippi River bank, it is exposed to water on one side and suburban area on the remaining sides. It is judged that the river is more than 5000 ft wide at the location of the building; it will be Exposure D for the building according to Section 6.5.6.3 of the Standard. The Exposure Category for components and cladding (C&C) Section 6.5.6.5 of the Standard states that, 'components and cladding design pressures for all buildings shall be based on exposure resulting in the highest wind loads for any direction at the site'. The reason for this requirement is that the values of external pressure coefficients for C&C given in the Standard were obtained in a wind tunnel independent of direction. Therefore, Exposure D is to be used in this example.

K_{zt} is the topography factor; it is defined in Section 6.5.7 in the Standard. As indicated in the Standard, this factor is applicable only when the site conditions and location of the building meet several conditions, including when the hill, ridge, or escarpment is isolated, the slope of the feature is greater than 6°, and the height of the feature is at least 15 ft for Exposure D. It is judged that the location of the example building does not meet the requirements as specified; hence, the topographic factor is not applicable, or $K_{zt} = 1.0$.

K_d is the wind directionality factor; it is specified in Table 6-4 of the Standard. There is a footnote in Table 6-4 of the Standard indicating that this factor shall only be applied when used in conjunction with load combinations specified in Sections 2.3 and 2.4 of the Standard. This footnote has caused concern among design professionals because load combinations related to wind loads have dead load as part of the load. It is argued that cladding subjected to wind loads does not have dead load or that dead load is negligible. Hence, the argument goes that load combinations in Sections 2.3 and 2.4 do not apply to cladding and that directionality factors do not apply to C&C. This argument is not correct. The intent of the Standard is to specify differentiation between load combinations for strength design and load combinations for allowable stress design procedures. The load factor is specified as 1.6 for wind load in load combinations for the strength design in Section 2.3 of the Standard, which is a reasonable value for both interior and coastal locations. The directionality factor accounts for a reduced likelihood that the maximum wind speed occurs in a direction that is most unfavorable for pressure. Also, it can be noted that Table 6.4 of the Standard specifies clearly that $K_d = 0.85$ for building components and cladding.

Summary to calculate velocity pressures:

K_z = varies with height z above ground using Exposure D; Table 6-3 of the Standard

K_{zt} = 1.0

Table 5.2 Velocity pressures, q_z

Height z (ft)	K_z	q_z (psf)
0–15	1.03	18.2
30	1.16	20.5
60	1.31	23.1
90	1.40	24.7
120	1.48	26.1
160	1.55	27.3
200	1.61	28.4
220	1.64	28.9

Note: value of K_h = 28.9 psf.

K_d = 0.85
V = 90 mph
I = 1.0

Using these values for parameters and Equation [5.2], velocity pressures are calculated and are shown in Table 5.2. The locations of heights at which the velocity pressures are calculated are chosen arbitrarily. It would be possible to choose different heights to determine velocity pressures and subsequently design pressures.

Prior to determining design pressures for C&C, mullions, and glazing panels in the present case, three items have to be determined: (1) enclosure classification of the building for internal pressure; (2) width of zone, *a*, for the pressure coefficient; and (3) effective wind area for each component and cladding.

1.   Enclosure classification governs the value of the internal pressure coefficient. There are three classifications: open building (not applicable in this case), partially enclosed building, and enclosed building. Professional judgment must be exercised to decide on the enclosure classification. The building under consideration is covered with glazing which, if properly designed and constructed, can resist design wind pressures. The location of the building is not in the wind-borne debris prone region. Also, since the building is located next to the river, it is judged that there will not be significant wind-borne debris that will break large numbers of glazing panels. The building is judged as an enclosed building. If the building is classified as a partially enclosed building the internal pressures would be much higher. Internal pressure coefficients are specified in Figure 6-5 of the Standard. Values of the internal pressure coefficient for the enclosed building are (GC_{pi}) = +0.18 and −0.18.

2.   Zone width, *a*, is defined as 10 % of the least horizontal dimension, but not less than 3 ft (the footnote in Figure 6-17 of the Standard). The

overall footprint of the building is 120 ft × 200 ft. The cutout in one corner of the building is 40 ft × 40 ft. This cutout is not likely to change significantly the overall pattern of wind flowing around the building. The least horizontal dimension of 120 ft will be used to determine the zone width a. The zone width a is applicable at every outside corner because there will be separation of wind flow creating high outward acting pressure at each corner. These corner zones (Zone 5) are shown in Fig. 5.6. The inside corner of the walls of the cutout of the building will not experience flow separation and will not see high outward acting pressures. Hence, it is not necessary to consider zone width a at that location. The zone width a signifies Zone 4 and Zone 5 on the walls; the zones are shown in Fig. 5.6. The width of corner Zone 5 is the larger of

$$a = 0.1 \times 120 = 12 \, \text{ft} \; (\text{controls})$$

or

$$a = 3 \, \text{ft}$$

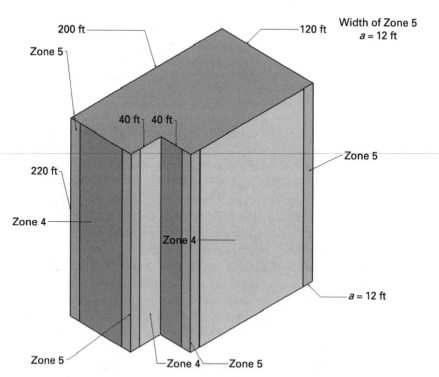

5.6 Example building showing wall corner zones.

3. The effective wind area, A, governs values of the pressure coefficients (GC_p). The definition of the effective wind area of C&C is the span length multiplied by the effective width that need not be less than one-third of the span length (per Section 6.2 of the Standard). Mullions are supported between the floors; hence the span length here is 11 ft. The glazing panels are assumed to be supported on all four sides; in this case it is reasonable to use the wind effective area as the tributary area. The mullions effective wind area is the larger of

$$A = 11 \times 6 = 66\,\text{ft}^2\,(\text{controls})$$

or

$$A = 11 \times (11/3) = 40.3\,\text{ft}^2$$

and glazing panels

$$A = 5.5 \times 6 = 33\,\text{ft}^2$$

The design pressures for C&C are obtained by Equation 6-23 of the Standard:

$$P = q(GC_p) - q_i(GC_{pi}) \qquad\qquad\qquad [5.3]$$

where

q $= q_z$ for the windward wall calculated at height z to be used with $(+\ GC_p)$

        $= q_h$ for the leeward and side walls calculated at roof height h to be used with $(-\ GC_p)$

q_i $= q_h$ for the enclosed building

(GC_p) = external pressure coefficient from Fig. 5.4

(GC_{pi}) = internal pressure coefficients of $+\ 0.18$ and $-\ 0.18$ as discussed above

For external pressure coefficients (GC_p), the log scale for the effective area in Fig. 5.4 has to be interpolated, which is difficult to do. However, it is possible to write equations for the lines in Fig. 5.4; equations for the lines in Figures 6-11 through 6-17 of the Standard are given in the ASCE Guide (Mehta and Delahay, 2002). Pertinent equations are shown here:

Wall Zones 4 and 5 for positive (GC_p):

$$(GC_p) = 1.1792 - 0.2146 \log A \qquad \text{for } 20 < A < 500$$

Wall Zone 4 for negative (GC_p)

$$(GC_p) = -1.0861 + 0.1431 \log A \qquad \text{for } 20 < A < 500$$

Table 5.3 Wall external pressure coefficients (GC_p)

	A (ft^2)	Zones 4 and 5	Zone 4	Zone 5
Component		($+GC_p$)	($-GC_p$)	($-GC_p$)
Mullion	66	+ 0.79	− 0.83	− 1.50
Glazing panel	33	+ 0.85	− 0.87	− 1.68

Wall Zone 5 for negative (GC_p)

$$(GC_p) = -2.5445 + 0.5723 \log A \qquad \text{for } 20 < A < 500$$

Using these equations and the effective wind areas calculated above, external pressure coefficients (GC_p) are determined for mullions and glazing panels; these values are shown in Table 5.3.

Design pressures can be positive or negative depending on the direction of wind and location of the wall component. The internal pressure can act toward (positive) the component surface and away (negative) from the surface. The controlling design pressure will be when external and internal pressures add to cause the maximum differential pressures. Sample calculations for controlling design pressures are given below; the controlling design pressures for mullions and glazing panels for the building are shown in the tables.

Controlling negative design pressure for mullions in Zone 5 of the wall:

$$\begin{aligned}
P &= q_h(-GC_p) - q_h(+GC_{pi}) \\
&= 28.9(-1.5) - 28.9(+0.18) \\
&= -48.6 \, \text{psf} \quad \text{(positive internal pressure controls)}
\end{aligned}$$

Controlling positive design pressure for mullions at height h = 90 ft in Zone 4:

$$\begin{aligned}
P &= q_z(+GC_p) - q_h(-GC_{pi}) \\
&= 24.7(+0.79) - 28.9(-0.18) \\
&= 24.7 \, \text{psf} \quad \text{(negative internal pressure controls)}
\end{aligned}$$

Design pressures for mullions are shown in Table 5.4 and for glazing panels are shown in Table 5.5. Maximum positive pressures (pressures acting toward the surface) and negative pressures (pressures acting away from the surface) are shown in the tables. These can be used for design purposes. If the cross-section of components, mullion or glass panel, are symmetrical the larger of the positive or negative pressures will control the design.

Table 5.4 Mullion design pressures (psf)

z (ft)	Zone 4		Zone 5	
	Positive	Negative	Positive	Negative
0–15	19.6	− 29.2	19.6	− 48.6
15–30	21.4	− 29.2	21.4	− 48.6
30–60	23.5	− 29.2	23.5	− 48.6
60–90	24.7	− 29.2	24.7	− 48.6
90–120	25.8	− 29.2	25.8	− 48.6
120–160	26.8	− 29.2	26.8	− 48.6
160–200	27.6	− 29.2	27.6	− 48.6
200–220	28.0	− 29.2	28.0	− 48.6

Table 5.5 Glazing panel design pressures (psf)

z (ft)	Zone 4		Zone 5	
	Positive	Negative	Positive	Negative
0–15	20.7	− 30.3	20.7	− 53.8
15–30	22.6	− 30.3	22.6	− 53.8
30–60	24.8	− 30.3	24.8	− 53.8
60–90	26.2	− 30.3	26.2	− 53.8
90–20	27.4	− 30.3	27.4	− 53.8
120–160	28.4	− 30.3	28.4	− 53.8
160–200	29.3	− 30.3	29.3	− 53.8
200–220	29.8	− 30.3	29.8	− 53.8

5.8   References

ANSI (1972) *Building Code Requirements for Minimum Design Loads in Buildings and Other Structures*, ANSI A58.1-1972, American National Standards Institute, New York, NY.

ANSI (1982) *Minimum Design Loads for Buildings and Other Structures*, A58.1-1982, American National Standards Institute, New York, NY.

ASCE (1990) *Minimum Design Loads for Buildings and Other Structures*, ASCE 7-88, American Society of Civil Engineers, New York, NY, 92 pp.

ASCE (1995) *Minimum Design Loads for Buildings and Other Structures*, ASCE 7-95, American Society of Civil Engineers, Reston, VA, 214 pp.

ASCE (2000) *Minimum Design Loads for Buildings and Other Structures*, ASCE 7-98, American Society of Civil Engineers, Reston, VA, 337 pp.

ASCE (2002) *Minimum Design Loads for Buildings and Others Structures*, SEI/ASCE 7-02, American Society of Civil Engineers, Reston, VA, 376 pp.

ASCE (2005) *Minimum Design Loads for Buildings and Others Structures*, ASCE/SEI 7-05, American Society of Civil Engineers, Reston, VA, 388 pp.

ICC (2000) *International Building Code 2000*, International Code Council, Falls Church, VA.

ICC (2006) *International Building Code 2006*, International Code Council, Falls Church, VA.

Mehta, K.C. and Delahay, J. (2004) *Guide to the Use of Wind Load Provisions of ASCE 7-02*, American Society of Civil Engineers, Reston, VA, 126 pp.

Mehta, K. C. and Marshall, R. D. (1997) *Guide to the Use of Wind Load Provisions of ASCE 7-95*, American Society of Civil Engineers, Reston, VA, 106 pp.

Morse, S. M., Norville, H. S., Mehta, K. C., Mc Donald, J. R. and Smith, D. A. (2005) *Wind Loads on Structures According to ASCE 7-05*, Standards Design Group, Lubbock, TX.

NFPA (2006) *NFPA 5000: Building Construction and Safety Code – 2006,* National Fire Protection Association, Quincy, MA.

Yeatts, B. B. and Mehta, K. C. (1993) Field study of internal pressures, in Proceedings of 7th US National Conference on *Wind Engineering*, UCLA, Los Angeles, CA, Volume 2, pp. 889–897.

6

Architectural glass to resist wind pressures

C . B A R R Y , Pilkington NA Inc., USA

Abstract: Successful glass design requires a detailed knowledge of expected or possible loads on the glass, their probability of occurring, and their duration. The expectations, both aesthetic and physical, of the end-user must be considered at this stage. A detailed stress analysis of a trial glass shape, type, and thickness will determine if a particular design is adequate. Repeat analyses of different glass types will allow selection of the most appropriate product for that particular situation. While glass is designed not to break under load, because it is a brittle material with a non-zero probability of breakage at most loads, the consequences of any such breakage must be fully evaluated before a successful design can be considered complete.

Key words: annealed, breakage probability, deflection, distortion, fully tempered, glass strength, heat strengthened, heat treated, laminated, post-breakage behavior, stress analysis, thermal stress, wind pressure.

6.1 Glass strength

Established practice in North America has used wind load glass charts based on a nominal probability of failure rate of 8 per 1000 at the design load. In simple terms this implies that if the maximum design wind load for a certain size and type of glass is applied, at any time in the life of the building, then one piece of glass of that size and type could be expected to break for approximately every 125 installed pieces. These charts are based on a conservative assessment of weathered glass strength and have demonstrated their conservative nature by the fact that although there have been wind storms where the maximum design wind velocity and pressure have occurred, there have been almost no recorded instances of properly designed, manufactured, and installed glass breaking at its design wind pressure. There is no known instance of breakage rates as high as the nominal 8 in 1000 from simple wind pressure (excluding tornadoes where ultimate wind pressures are not known or recorded due to extensive property destruction and damage to local anemometers) with properly designed glass. Note that these comments do not address glass breakage caused by wind created debris impacts. An added factor of conservatism in window glass design is created from the fact that glass is only available in a finite number of thicknesses. This means that a glass designer must select the next available nominal glass thickness that is greater than the theoretical one dictated by the strength design procedure to meet the actual load specified.

In an overload situation glass does not break all at once across a plate; a fracture will originate at one point and grow, rapidly or slowly, depending on other factors such as the magnitude of the applied tensile stress, perhaps in a single crack or branching into multiple fractures. A brittle fracture begins at the point where the combination of a weaker section of glass containing a small blemish or stress concentrating scratch combined with a relatively large tensile stress results in the most probable point of origin.

The tensile strength of pure glass, with perfect surfaces, is about 1 000 000 psi (6.7 GPa). This value can be obtained by tensile testing freshly made glass fibers, which have never been in contact with any materials, in a vacuum. However, when normal surface imperfections, even though they can be small enough to be invisible to the naked eye, are included, the tensile strength of ordinary, annealed, in-service glass is greatly reduced to a value in the range of 3000 to 9000 psi (21 to 62 MPa) for short duration loads. Glass is therefore designed to have an acceptable probability of surviving the worst expected loads at a specific building location, while good design practice will also consider the resulting behavior of broken glass, under load, in the unlikely probability of it actually failing at its specified load; i.e. what would happen if pieces should fall out of the frame, or will the integrity of the building interior be seriously compromised by exposure to the weather?

The behavior of cracked or shattered glass depends largely on the nature of the load applied after glass breakage occurs; glass that breaks from a wind or snow load during a storm is likely to be further subjected to the design load, before the storm passes. In these cases the ability of cracked glass to resist the continuing load is severely compromised and pieces could be expected to fall out of the frame. However, glass that has cracked from excessive thermal stress or from a single impact of a low mass body (a stone from a slingshot, for example) will readily remain in the frame provided the cracked pieces all have some part of their perimeter connected to the glazing frame. Typical solar control glazing incurs thermally induced stresses on a daily basis, so a damaged piece of glass can be expected to show crack growth at the fracture tips over time. Howerver, this thermal crack growth only occurs where there is a locally induced stress from a temperature gradient in the glass. If the glass gets uniformly hot in the sun, and can expand uniformly, it will not be stressed. Often it will take many months of normal weather exposure for some cracks to advance visibly.

Pure architectural glass, which is usually soda-lime glass, is extremely strong, but the brittle nature of glass means that a small blemish under a much smaller stress can become the origin of a fracture. The probability of finding such a blemish increases as the stressed area of a plate increases, even though the magnitude of the stress remains constant. This results in an 'area effect' whereby larger plates, with greater areas under the same stress, are considered more likely to break. The ASTM E1300 glass load resistance charts[1] allow for this effect by means of the failure prediction model (FPM) method of computation[2].

The slow growth of a crack under constant stress, and in the presence of water vapor, has been shown to be a form of water-based stress corrosion, where atomic silica bonds are broken and the crack enlarges by a few molecules at a time. The chemistry of this process is well described by Michalske and Bunker[3]. This is the explanation of the 'static fatigue' effect, which effectively halves the strength of glass under long-term (weeks) snow load as compared to the short duration (seconds) load of a wind gust. The strength of annealed glass is considered to be proportional to the inverse of the load duration raised to the power of about one over 15 or 16: $R_t = R_{ref} t^{-1/n}$, where R_t is the average resistance under a constant pressure for time t (minutes), R_{ref} is the reference resistance under a sustained pressure for unit time (1 minute), and n is an exponent value of 15 (see CAN/CGSB-12.20-M89 [4] and Johar[5]).

Heat-strengthened (HS) and fully tempered (FT) glasses are nominally twice and four times stronger than annealed glass (AN), respectively. These thermally toughened glass types are made by heating the glass so that it becomes soft, and then rapidly quenching it to freeze the skin and put it into compressive stress, with a corresponding tensile stress in the glass core, while the glass cools.

As glass only breaks under tensile stress it is difficult to calculate the strength of (HS) or (FT) glass accurately because the compressive stress in the skin must first be overcome. This difficulty arises from the nonuniformity of the quenching process caused by the air jets from many different nozzles creating a nonuniform air flow at the glass surface, resulting in a variation in the surface compressive stress level across the plate. With the tensile stress core protected from water and water vapor, static fatigue effects are less pronounced with heat-treated glass.

The importance of accurately determining the wind gust duration can thus be appreciated if the variability in the glass strength values is to be acceptable. When the ASTM E1300 revised glass load charts were re-published in 2004 they were adjusted to accommodate the revised ASCE 7[6] changes in maximum wind speed duration: the 60 seconds duration original values were reduced to 3 seconds, with a gust duration of 1 to 10 seconds. These shorter gust duration times resulted in higher average gust speeds.

While the static fatigue effect of load duration on glass can be readily evaluated with the exponential formula given above, the time interval used to average the wind speeds and pressures is not directly related mathematically to the duration of the peak load. Nevertheless, the net result of increasing the maximum wind gust load and reducing the gust duration for the load resistance charts meant that the glazing industry saw little change in the glass type and thickness actually used to glaze a given size opening because the increase in peak wind load closely matched the corresponding increase in apparent glass strength due to the shorter load duration.

Artificial weathering test results, according to proposed European EN standards, show a reduction in strength when comparing the aged product to as-manufactured new glass. The resulting strength value of severely aged glass is finite and does not continue reducing with further weathering. This concept of a finite residual strength for continuously weathered glass is confirmed by the fact that there are no records of 'old' (ages of 50 or 100 years and more) glass in buildings breaking any more frequently than 'new' glass from high-velocity wind loads. The conclusion is that weathering of glass in service reduces its strength to a finite level, and further weathering does not reduce the strength any more. This demonstrated wind load resistance of installed glass of many years age has confirmed the reality of the theoretical self-repair mechanism, whereby the stress concentrating crack tips at in-service damage points are rounded out and relieved at the atomic level by water-based corrosion during the no-stress periods between windstorms. This allows the long-term tensile strength of soda-lime glass to stabilize at some useful design value around 3000 psi (21 MPa), even though it is far less than the failure strength of the original 'new', unblemished, plates, which can be 20000 psi (140 MPa) or higher. The ASTM E1300 standard is based on data from destructive tests of in-service, aged glass samples that include the weathering effects as detailed above.

Sloped glazing receives short duration wind loads acting normal to the glass surface and long duration snow loads acting vertically, as well as the permanent, vertical acting, load from self-weight. This analysis is even more complicated with sloped insulating double glazing due to load sharing between the glass lights when the double glazed air space is sealed. When the lights are of differing thickness, the thinner, more flexible glass transfers most of its load to the thicker, stiffer glass. Methods exist in the ASTM E1300 standard to combine these different loads into one equivalent lateral load.

The short duration of a maximum design wind gust is recognized in the higher published load resistance values of glass for wind loads (of nominal 3 second duration) as compared to the lesser resistance values shown for longer term snow loads, or the permanent loads of aquarium windows. Performance requirements of an aquarium window, especially when considering the consequences of an unanticipated breakage condition, usually require the use of multiple layers of annealed glass in a laminated product. Such windows are often designed with annealed glass layers to a maximum surface tensile stress of only 1000 psi (6.9 MPa). The selection of relatively thick, half inch (12 mm) to one inch (25 mm), annealed glass for aquarium viewing ports means that, in the event of breakage, the glass is more likely to fracture into only a few large pieces that will usually partly remain wedged in the glazed opening, even if the glass is single layer monolithic (not laminated). This post-breakage behavior inhibits the escape of water and fish, while reducing the risk to viewers.

It should now be clear that while window glass can be designed to resist specified and finite loads with a given probability of success, there is no glass type that will reliably break at a certain precise load, as would be required in the case of a pressure relief rupture diaphragm, for example.

6.2   Stress analysis

Window glass is mostly rectangular in shape and is typically supported in stiff frames along all four edges. The glass edges are presumed to be 'simply supported' in that they are free to rotate with the line of support acting as the axis of rotation, or move in the direction of the plane of the glass, but they do not move perpendicular to that plane. The frames are assumed to be stiff enough so that they do not deflect, under full load, more than 1/175 of their length. This maximum deflection limit is set by the boundary conditions in the analysis used by the failure prediction model computer program that generated the glass strength charts in ASTM E1300.

When a wind load is uniformly applied to glass supported on all four edges, the glass will first bend and deflect at the center of the pane, in proportion to the load, until the deflection is equal to about half its thickness. After that, membrane stresses are generated as the glass tries to stretch to take up a domed shape in the central area, with resistance to the stretching being supplied by the glass near the perimeter, which takes a more conical shape because of the frame. This membrane effect stiffens the glass considerably. The stress function from membrane and bending effects then becomes nonlinear with respect to both load and deflection, and thus requires a finite difference or a finite element analysis method for proper evaluation.

With four-edge supported, square, or rectangular shapes of less than about 2:1 (height: width) aspect ratio, the maximum stress occurs on the diagonals near, but not at, the corners. For higher aspect ratios the

6.1 Gable end shape.

maximum stress occurs along the middle of the long axis. Thus, the cut edge quality of the glass is not as critical a factor in the ultimate strength as might be expected. However, for rectangular glass plates supported only along two opposite edges, the maximum bending stress from wind loads will occur at the free edges and so the cut edge quality becomes critical. The edge quality of most window glass is also important because thermal stresses, from heating and cooling conditions, can create significant tensile stresses at the glass edges.

Nonrectangular shapes, circles, triangles, trapezoids, etc., should be analyzed for maximum stress with finite element methods if they are loaded to near their capacity. A first simplified analysis is often made using a piece the size of a bounding rectangle with the available load resistance charts for rectangular shapes. This simplified analysis is not always conservative as in the case where two 45° corners have been cut from a 2 to 1 aspect ratio rectangle to create a gable end shape (Fig. 6.1). In this case there is an increase in stress, under uniform load, of around 10 or 20% depending on glass thickness and glazing details, at the 135° corner as compared to the maximum stress in the larger, simple rectangle.

For thermal stress considerations with the maximum stress in the glass edges, away from the corners, the cut edge quality is critical. With today's thermally efficient glasses using spectrally selective compositions and coatings to control unwanted solar heat gain, and with the addition of low emissivity coatings to further improve thermal performance, the resulting thermal stress will often require the use of heat-treated glass to prevent breakage. A thermal stress analysis needs to be made, independently of the wind load analysis. Typically, the maximum thermal stress situation for the exterior light of double glazing occurs in still air conditions when heat loss by natural convection to the exterior is at a minimum. Thermal stress is essentially a dynamic function usually caused by a rising sun, or the sun appearing from behind a cloud, quickly heating the exposed areas of glass while the edges, shaded within the frame or partly shaded by frame projections or other materials, remain relatively cool. After some time the exposed area will reach an equilibrium temperature while the shaded glass edges continue to slowly become warmer due to heat being conducted through the frame and from the exposed areas. The durations of such maximum transient thermal stresses have not been accurately computed, but it is reasonable to expect them to last considerably longer than the 3 second wind of the current load resistance charts.

Proprietary programs are offered by major glass manufacturers to assist their customers with thermal stress analysis of combinations of their particular glass products in unique glazing installations. A work committee at ASTM is currently in the early stages of developing a comprehensive thermal stress analysis standard. This committee recently published a preliminary version for single glazing[7].

The edge strength of a glass plate is weaker than that of the surfaces away from the edges. While the aged, in-service, surface strength of glass has been frequently measured using simple ring-on-ring concentric pairs of load applicators, there is greater variability in the failure strength or load resistance of a cut edge. Beam samples can be measured with a four-point bending rig to create constant bending stresses between the inner loading points. With this device the surface stress extends to the glass edges. However, glass cutting is essentially a controlled damage mechanism that allows a separation plane to open or snap, along a relatively weak scored line when the plate is stressed by bending, or sometimes by differential thermal heating and cooling. The amount of residual damage at the glass edge after snapping is a function of the quality of the scoring and snapping processes. The snapping process can create edge damage if it does not create a pure bending stress at the score line location or if it allows the newly formed edges to contact each other during the separation phase.

Relatively weak glass edges can be strengthened somewhat by fully polishing them to remove all visible imperfections, but it is not possible to see if the polishing process has removed the very tips of all the original imperfections. Fine seaming with an abrasive belt can be used to remove some edge damage, but the action of creating the fine seamed surface actually adds further imperfections. Thus, edge seaming will typically strengthen weaker specimens, but can also weaken stronger ones. ASTM E1300 gives some approximate edge strength values, but actual sample testing would be required to obtain more accurate results for particular edge conditions.

Before finite element analysis was readily available it was assumed that the load resistance of a given thickness glass was inversely proportional to the glass area, and aspect ratio effects were ignored. Glass strength charts were published, with straight lines for different glass thicknesses, for allowable load versus area, both in logarithmic scales (Fig. 6.2).

The current ASTM E1300 charts have 'S'-shaped load lines on a linear

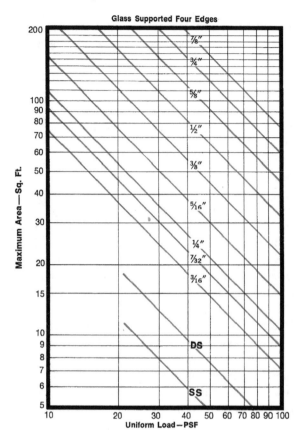

6.2 Old style straight line log–log load versus area chart (reproduced from *Glass for construction*, Libbey–Owens–Ford, 1974[8], by permission of Pilkington).

scale, length-by-width chart, which reveal that aspect ratio effects have been accounted for, as shown in Fig. 6.3 in the single glazed chart for $\frac{1}{4}$ in. (6 mm) annealed glass, supported on all four edges, under a 3 second duration uniform lateral load. The inflection, or 'S' shape, in the load line indicates how the location of the maximum stressed area, and probable origin of a fracture in an overload condition, changes from being near the corner to being near the middle of the short span, near the central area, as the aspect ratio of the plate increases; square plates have less load capacity as their size and area increases, but the load capacity of high aspect ratio plates does not change as much when their long dimension (and area) increases. The strength reduction as the long dimension increases is a measure of the increasing probability of encountering a severe enough blemish to act as a glass fracture origin even though the level of stress does not increase significantly.

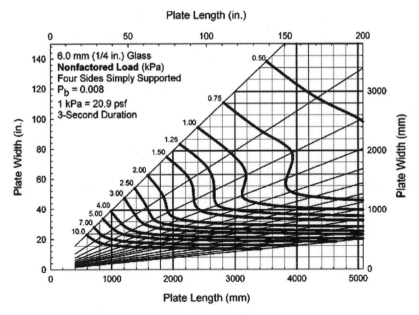

6.3 ASTM E1300 chart for 6 mm ($\frac{1}{4}$ in) glass load resistance.

For rectangular glass supported at one edge, or on two or three edges, the maximum stress occurs at a free glass edge and so the cut edge quality of the glass becomes critical. Glass load charts usually recognize that glass edges are typically weaker than glass surfaces and so the chart values are adjusted accordingly. ASTM E1300 has load charts for these three conditions and also gives suggested design values for stressed edges. Glass analysis with other support conditions can then be made using the linear load/stress/ deflection relationships obtainable from standard engineering texts such as *Formulas for Stress and Strain* by Roark and Young[9].

The popular point supported, flush glazed systems, which have tempered glass retained by structural bolts through drilled holes near the glass corners and at intervals along long edges, is a specialized subject. It requires particularly careful analysis to account for all the degrees of freedom of translation and rotation at each support point, as well as performing an accurate stress analysis at the hole itself, plus quantifying the allowable design stress for the drilled hole surface. There are a number of papers on this subject available, for example, from the biennial Glass Processing Days conferences held in Tampere, Finland. (see www.glassfiles.com for details[10]).

North American windows typically have two glass lights in a sealed double glazing (insulating glass, or 'IG') unit, for energy efficiency. The stiffness to compression or rarefaction of the air in the sealed air space compared to the relative flexibility of the glass means that for a normal size window the two lights of glass in a sealed unit will essentially move in parallel, by the same amount, under a uniform positive or negative wind pressure load. Ideally, this gives a 50/50 load sharing effect with both lights being equally stressed, in effect doubling the wind load capacity of double glazing as compared to single glazing with the same glass thickness. This load sharing effect needs to be reduced slightly to allow for the out-of-parallelism of the two lights caused by barometric pressure, temperature, and small altitude differences between the time and place of manufacture and installation. This load sharing is very effective when the glass size is relatively large and thin. Conversely, when the short dimension of rectangular glass is less than about 20 in (500 mm), the glass is $\frac{1}{8}$ in (3 mm) or thicker, and the sealed air space is $\frac{1}{2}$ in (12 mm) or more, changing air space pressures will cause large stresses in the glass. It is seen in practice that for properly made and properly installed insulating glass units with flexible edge seals, where the glass edges are free to rotate slightly when the center of glass deflects, then the largest total sealed air space that will not cause glass overstressing in normal changing weather conditions is about 1 in (25 mm). A triple glazed sealed unit, with two air spaces each of $\frac{1}{2}$ in (12 mm) depth, can be conservatively assumed equivalent to a double glazed unit with a 1 in sealed air space for terms of load sharing and weather related glass stress from air space pressure.

If a double glazed IG unit is made of two lights of differing thickness or stiffness, then the stiffer glass will carry more of the load. As the stiffness of glass in simple bending is proportional to the cube of its thickness, doubling the thickness of one light will make it eight times stiffer than the other one, so the thicker light then carries eight times the load. A typical example of asymmetric load sharing is in an IG skylight with tempered outer glass of $\frac{1}{4}$ in (6 mm) and a laminated, annealed inner light of two plies of $\frac{1}{8}$ in (3 mm): under a long-term snow load, when the laminated light behaves like two layers sliding on each other, the outer $\frac{1}{4}$ in (6 mm) light is carrying four times the load of the inner laminated glass plies. There is a further subtlety to this example: under a short duration load, laminated glass can behave in a monolithic manner (i.e. its strength and stiffness in bending will be equivalent to a monolithic glass plate of the same total glass thickness), especially if the interlayer is relatively cool and stiff, i.e. at room temperature or below. Therefore, this skylight example will also need to have both lights checked for the short duration load situation where the two lights will be carrying equal loads.

In IG design it is considered that a double glazed unit has four planar surfaces, apparently doubling the chance of finding a critical blemish from which a crack could start, as compared to single glazing with two such surfaces. However, in a sealed IG, two of the four surfaces are in a dry, protected atmosphere and are not subjected to normal, in-service, humidity based, stress corrosion or aging. Thus, the load resistance of an IG unit will appear to be slightly weaker than twice that of single glazing of the same size and glass thickness, when the sealed air space pressure effects are taken into account.

The US standard, ASTM E1300, as referenced by the US ICC Building Codes[11], takes all the above factors into account in its procedures. The load resistance charts can provide values for short duration wind loads as well as long duration snow loads, for single or double glazing.

The Canadian glass strength standard, CGSB 12.20, computes the load resistance of glass by a different 'limit states design' method that includes 'material factors' and 'importance factors'. Also, the wind loads are not calculated in exactly the same manner as in the US model building code. CGSB 12.20 does include the load duration and area effects described above. However, the end results are very similar to those given by ASTM E1300.

Committees at European (CE) and International (ISO) organizations are developing glass strength standards on similar lines. It is planned to base them on a body of data for artificially weathered glass by using sand grains, trickled from a height on to many sloping glass samples, to simulate the aging effect of installed glass in service. Finite element studies and other stress calculation formulas have been used to compute resulting stress values for different glass thickness, shape, and support conditions under uniform load. It is hoped that the resulting design loads for glass will have a more realistic breakage probability value than the overly conservative 8 in 1000 number associated with the North American ASTM E1300 method.

6.3   Glass types

Annealed glass is the standard float glass product that has been slowly cooled after forming in the molten tin float bath. The slow, uniform cooling to room temperature results in a relatively stress-free material that can be cut, drilled, edge worked, etc. Heat-strengthened (HS) glass is nominally twice as resistant to uniform wind loads as standard annealed glass, and fully tempered (FT) glass is nominally four times as resistant as annealed glass. These factors are conservative because they are based on the minimum values of compressive stress that must be exceeded if heat-treated glass is to meet the ASTM C1048 standard[12].

The heat treatment processes for HS and FT glass involve heating the glass until it becomes soft and then uniformly quenching it on both sides with powerful air jets to cool and solidify the outer skin rapidly. The inner core of the glass then cools, shrinks, and puts the skin into a state of compression, with an equal and opposite tensile stress in the almost flawless middle core of the glass thickness. The quench process for HS glass is less vigorous than for FT glass and so creates less compressive stress on the

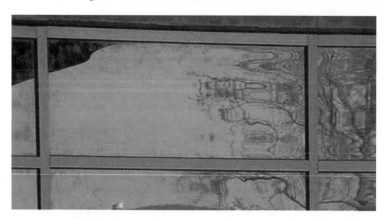

6.4 Typical reflection distortion in heat-treated glass for relatively long viewing distances.

exterior surfaces. Because glass breaks primarily under tensile stress, any wind load that causes bending must first overcome the built-in compressive stress of the heat treatment process, and so heat-treated glass is significantly stronger than annealed glass, which has essentially no built-in surface compressive stress. The stress corrosion process that causes an apparent static fatigue in annealed glass, as demonstrated by its lessened load resistance to long duration loads in the presence of water vapor, does not occur with the compressive stressed surfaces of heat-treated glass and so the reduction in strength for long duration loads is less apparent in their load resistance charts.

Because heat-treated glass (HS and FT) has had its temperature raised to the point where the glass becomes soft, it will not be as flat as annealed glass and will often show some visible distortion, especially in reflected images when viewed at longer distances, as compared to annealed glass (see Fig. 6.4).

Heat-treated glass can also have visible optical effects when illuminated with polarized light, such as that occurring on a sunny, blue-sky day. Stressed glass rotates polarized light and under certain conditions can create constructive and destructive interference effects resulting in light and dark visible blotches corresponding to areas of greater or lesser stress. This condition is visible, to the naked eye, to some degree in all heat-treated glass under appropriate lighting conditions and is a normal byproduct of the heat treatment process, as shown in Fig. 6.5.

When broken, HS glass will have a break pattern of relatively large pieces, similar to annealed glass, while FT glass shatters into myriad cubes each about the size of the glass thickness. Edge working (polishing or seaming), beveling, hole drilling, vee grooving, sand blasting, etc., must be carried out

6.5 Visible quench pattern in laminated heat-treated glass reflecting a uniform blue sky.

on the annealed glass piece before any heat treatment, HS or FT, or chemical tempering is performed. Surface treatment of any type that penetrates the compressive skin of a heat-treated product can only reduce its strength, usually by some unknowable amount, and so must be avoided.

The large, locked-in stresses of heat-treated glass can, on occasion, create an unusual problem of spontaneous breakage if a very small (typically 0.005 in (0.1 mm) diameter) nickel sulfide inclusion should be in the glass. These inclusions would normally grow in size by about 4% during the annealing process, when the glass is still in its plastic state, and so are harmless in annealed glass. However, the rapid quench of the HS and FT heat treatment process can freeze the inclusion in its smaller state. Over a number of years it can increase in size, by about 4%, as it changes its crystalline structure, thus increasing the tensile stress within the surrounding glass and possibly causing breakage without any other load being applied. Spontaneous breakage is less likely to occur with HS glass than with FT glass because of the lower stress levels in HS glass. A secondary 'heat soaking' process of holding the heat-treated glass in a hot oven at a few hundred degrees F for a few hours can reduce, but not completely eliminate, spontaneous glass breakage. As it is impossible to guarantee the elimination of such contamination during float glass production, and because the small size of nickel sulfide inclusions is below the detection limit of automatic fault inspection equipment, all tempered glass designs should allow for this unlikely breakage possibility. For example, most building codes do not allow sloped, tempered glass higher than 12 ft over occupied areas where broken pieces could fall to the floor. For sloped glazing, tempered glass can be the uppermost light in a double glazed unit, or laminated to another ply, to give good resistance to falling objects or hailstones and the lower light can be laminated to resist fallout, even if broken, when properly glazed. Vertically glazed heat-treated glass is not subject to this restriction because, when broken, the pieces will typically remain in the frame for some time, wedged together, unless a large lateral load is subsequently applied.

Laminated glass is made by assembling a sandwich of two or more plies of equal or differing layers of glass with a transparent adhesive interlayer. This interlayer, usually polyvinylbutyral (PVB), a proprietary ionomer, or epoxy between the two plies of glass, has nearly the same strength and stiffness as monolithic glass under short duration loads, but acts as a 'safety glass' when broken by remaining in the frame and offers significant penetration resistance. The plastic interlayers are available in clear, transparent or diffuse, and tinted materials. The PVB or ionomer materials are bonded between two glass plies by heat and pressure in an autoclave. Epoxy interlayers are cold liquids poured into a framed space between two glass plies. The curing and setting of the epoxy can be accelerated with heat or UV light. The plastic and epoxy interlayer materials have similar indices of refraction as glass, so their presence cannot be visually detected at the glass-to-plastic or glass-to-epoxy interfaces.

The uniform load resistance of laminated glass is difficult to compute exactly. The plastic interlayer materials have a stiffness under short-term loads, especially at room temperatures and lower, which make the glass behave in a monolithic manner under short duration loads. It is not unreasonable to use wind load charts for a monolithic glass of the same thickness as the total of the two glass plies, ignoring the plastic interlayer. For long duration loads or at high temperatures, a more conservative method is to use a layered approach, which assumes that each ply carries half the load (assuming they are of equal thickness) with no shear stress resistance offered by the interlayer. The ASTM E1300 standard includes charts for laminated annealed glass. A recent proposal for ASTM E1300 plans to quantify the interlayer shear resistance, and hence allow calculation of the added load resistance at different temperatures by using specific values for the dynamic shear modulus of the interlayer, as supplied by the material manufacturer.

The main reason for using laminated glass is usually to supply protection to the building envelope against penetration, and so the important variable then becomes the load resistance of the interlayer material itself after the glass plies have broken. If needed, this value is best obtained by full-scale testing. Dynamic tests with different sized debris impacts are designed to simulate hurricane conditions. There are other ASTM tests designed to simulate the resistance of glass to differing forced entry conditions.

6.6 Example of transmission distortion caused by laminating plies of heat-treated glass.

Laminated FT or HS glass is the ultimate product in terms of wind load and impact resistance and in building envelope protection. However, even here, there is a disadvantage: the optical quality of transmitted light will be diminished by having two plies of heat-treated glass, which cannot be as flat as the original annealed pieces and so will create overall thickness variations when laminated. These small variations will result in positive and negative lensing effects, typically only visible when the viewer-to-glass and glass-to-viewed object distances are relatively long (see Fig. 6.6), or when the viewing is at a high incidence angle to the glass plane. This type of distortion is less visible when thicker plies (6 mm or greater) are used in the laminate.

6.4   Deflection

Window glass deflection is an important calculation in the design process. There are no code limits on deflection, other than that of a free glass edge that must not deflect more than its glass thickness, relative to a nonmoving adjacent glass edge. This situation is often encountered in interior shopping malls where tall lights of single glass are only framed on the top and bottom edges. The IRC building code addresses the possibility of such a finger-pinching opening developing if one light were loaded by, say, a crowd of people pushing laterally against the glass. A horizontal handrail lateral loading of 50 lb/lineal ft (729 N/m) at 42 in (1.07 m) above the walking surface is often used as a design load in this situation. A value of 5 psf (0.24 kPa) is also often used to model a malfunctioning HVAC (heating ventilation, and air conditioning) system and represents a worst case uniform load for these interior installations.

The most important issue with deflection is that the glass must not deflect sufficiently to allow it to become free of its framing. Annealed glass may fracture at relatively small deflections, but fully tempered glass is very unlikely to break while bending under pure wind load. The tempering process increases the strength of glass by a factor of four with no change to the modulus of elasticity or the bending stiffness. Large, fully tempered sizes must, therefore, be checked for deflection at design loads to ensure that a glass edge will not pull out of its frame, even partially. A simplified analysis can be made assuming that four-side supported glass bends into a cylindrical curve, with its axis parallel to the long edge. Calculating the difference between the chord length and the arc length (which is unchanged from the original, undeflected, flat glass dimension) will give the worst case pullout dimension, assuming there is no in-plane movement at the other edge.

Glass deflections are much greater when the rectangular panel is supported only at two opposite edges. For example, a 100 in (2.54 m) high light of $\frac{3}{8}$ in (10 mm) glass (0.355 in (9.02 mm) ASTM C1036 standard minimum thickness for $\frac{3}{8}$ in (10 mm) nominal glass), single glazed, simply supported only at top and bottom edges, will deflect 3.5 in (89 mm) under a uniform load of 15 psf (0.7 kPa). This leads to a worst case condition where the top edge pulls down by 0.33 in (8 mm), assuming that the weight of the glass prevents the bottom edge from moving vertically upwards. To perform this calculation one first gets the center-of-glass deflection (d) under the design load from ASTM E1300 or the window glass design program. The formula is PO $\approx 2.67 \times d^2/S$, where PO = Pullout, d = center-of-glass deflection, and S = span or bent length of glass, and is reasonably accurate for most window sizes.

The Pilkington publication *Good Glazing Guidelines*[13] recommends that $\frac{3}{8}$ in (10 mm) glass be glazed with a $\frac{7}{16}$ in (11 mm) edge bite into the frame. Reducing this edge bite dimension to only $\frac{1}{8}$ in (3 mm) under load, as in example Fig. 6.7, is obviously not good practice and creates the risk of the frame lip bending and damaging or releasing the glass edge. In a four-edge supported glazing panel undergoing large deflections the mid section of an edge may come partially free from the glazing pocket and will invariably be crushed against the frame edge and break as it tries to flex back into place after the lateral load has subsided.

Aesthetic considerations create limits to large deflections in glass. The general public is not familiar or comfortable with the concept of glass visibly flexing and deflecting in a high windstorm – even though this does not necessarily constitute any strength limits being exceeded. For this reason a nominal maximum deflection limit of 0.75 in (19 mm) is often used for design purposes. Earlier editions of the ASTM E1300 standard had a dotted line on the uniform load graphs, indicating when a particular combination of glass size, thickness, aspect ratio, and load resulted in a center-of-glass

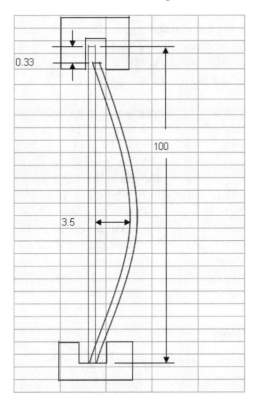

6.7 Pullout from a frame caused by deflection of two-side supported glass (dimensions in inches).

deflection of $\frac{3}{4}$ in (19 mm) or greater. This was simply a guideline for information purposes only, but it was removed in the 2004 edition because it was too often interpreted as being a mandatory maximum limit. It should be recognized that glass deflections are perceived more by the distortion of reflected images as the glass bends and flexes out of plane, rather than by actual direct perception of glass movement normal to its original plane.

A frequently encountered design concept is the 'wall of glass', with no structural members obstructing the view. This concept is seldom achievable without major innovations. Relative to steel, for example, glass is fairly flexible, with a modulus of elasticity comparable to that of aluminum. Fortunately, the bending stiffness of a glass beam is proportional to the cube of its thickness, so doubling the thickness of a glass pane increases its bending stiffness by a factor of 8. Glass is available in thicknesses up to 1 in (25 mm), but the thicker glasses are not available in all tints or with all coatings.

Insulating glass (IG), which is typically required to meet HVAC requirements in modern buildings, is not well suited to being installed with unsupported edges as in a two-sided support glazing with the other two opposing edges free to deflect. The resulting shear stress from wind load in the IG sealant needs to be evaluated and controlled at large glass deflections to prevent tearing of the seal and consequent failure and fogging of the IG unit.

A reasonable method of stiffening a single glazed glass wall to resist wind loads, while avoiding the use of vertical framing members, is to clamp the top and bottom glass edges instead of using the simply supported edge category of a normal glazing frame. A clamped lower edge, similar to that used at the bottom edge of glass handrails, will effectively stiffen the lower part of the light. If the upper framing is to be similarly stiff against rotation it must also include a vertical sliding detail to allow for expansion, contraction, and creep movements of the glazing system in the vertical plane. Such a system of two-side support with clamped edges top and bottom to resist wind load will create bending moments at top and bottom edges that will reduce the maximum stress of a simply supported glass light by 33% and will reduce the center-of-glass deflection by a factor of 5.

6.5   Design procedure

The ASTM standard E1300, *Standard Practice for Determining Load Resistance of Glass in Buildings*, is referenced in the US *International Building Code* (IBC) as the method for determining the glass type and thickness to withstand the wind load determined from ASCE 7, *Minimum Design Loads for Buildings and Other Structures*. The procedure in E1300 is lengthy: it starts with obtaining a nonfactored load resistance value from a chart for a particular glass thickness and edge support condition, when the plate length and width are known. If the glass is laminated, heat treated or subject to long duration loads then a glass type factor, chosen from a series of tables, is applied. If the glass is in a sealed double glazed unit (insulating glass unit) then another factor, chosen from another series of tables, is applied to allow for load sharing effects between the lights. The resulting overall load resistance of the insulating glass unit is then compared to the specified design load to determine if the chosen design is adequate.

A simple example in SI units is given below for single glazed annealed glass under a short duration wind load using the E1300 load resistance charts.

Example. Determine the nonfactored short duration load for a 1500 by 1200 by 6 mm monolithic annealed glass plate.

1.   Enter the horizontal axis of the nonfactored 6 mm load chart for four-side supported glass at 1500 mm and project a vertical line (see Fig. 6.8).

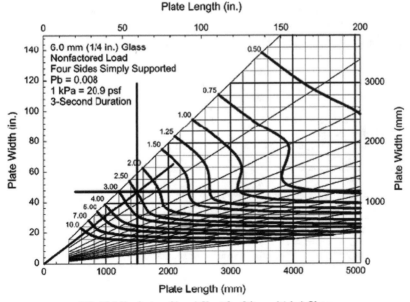

FIG. A2.1 Nonfactored Load Chart for 6.0 mm (¼ in.) Glass

6.8 Nonfactored load for ¼ in. (6.0 mm) four-sided supported glass.

2. Enter the vertical axis of the nonfactored load chart at 1200 mm and project a horizontal line.
3. Draw a line of constant aspect ratio through the intersection of the lines described in steps 1 and 2, and through the graph origin, and interpolate along this line to determine the nonfactored load. The nonfactored load is thus found to be 2.5 kPa. This is the short duration load resistance for a 1500 × 1200 × 6 mm rectangular plate, simply supported on all four edges, for a probability of breakage of less than 8 in 1000.

A more complex example, in inch-pound units, examines a nearly horizontal glass skylight, size 4 ft × 5 ft, required to resist a 30 psf short duration wind load and a 45 psf long duration snow load. Table 6.1 shows a series of ASTM E1300 derived load resistance results, along with the design procedure steps, used to select an appropriate glass type. Note that the procedure includes the stress and deflection from the self-weight of the sloped glass.

These procedures can be executed more conveniently and more accurately by using readily available computer programs such as *Window Glass Design* from Standards Design Group (see www.standardsdesign.com[14]). This program uses the same methods and algorithms as in ASTM E1300 and can produce the same answer in a printed form.

Table 6.1 ASTM E1300 derived load resistance results and design procedure steps used to select an appropriate glass type

Glass type number	Glass type description	Short duration load resistance (psf)	Center-of-glass deflection (in) short duration load	Long duration load resistance (psf)	Center-of-glass deflection long duration load (in)
1	6 mm annealed (AN) single glazed	51.4	0.43	25.7	0.55

Glass type 1 does not meet the long duration load specification. Try a stronger glass (HS).

| 2 | 6 mm HS | 103 | 0.43 | 66.8 | 0.55 |

Glass type 2 will not be suitable for overhead glazing in most building codes. Try annealed laminated glass.

| 3 | 3 mm AN + 0.76 mm PVB + 3 mm AN | 51.6 | 0.5 | 25.8 | 0.61 |

Type 3 does not meet the long duration load specification. Try heat-strengthened laminated glass.

| 4 | 3 mm HS + 0.76 mm PVB + 3 mm HS | 103 | 0.5 | 67 | 0.61 |

Type 4 meets the specified loads but will possibly have some visible transmitted distortion. The deflection of 0.61 in might cause water ponding issues if the glass is nearly horizontal. Try an insulating glass unit.

| 5 | 6 mm FT / air / 3 mm AN + 0.76 mm PVB + 3 mm AN | 103 | 0.36 | 115 | 0.48 |

Type 5 meets the specified loads, most building code requirements, and normal optical expectations. It will also have a higher R value than single thickness glass panels, which will reduce winter heat loss and help prevent cold weather condensation on the room side surface.

6.6   Post-breakage behavior

Having demonstrated the difficulty in arriving at a precise glass design, due to uncertainties involved in accurately defining the wind load and in accurately determining the glass strength, it becomes necessary to consider the possible consequences of any unintended glass breakage that may occur in the normal lifetime of a building.

When annealed glass breaks it typically has a few fully developed cracks and most, if not all, of the pieces have one or more edges connected to the frame. If a viscous mastic or curing-type glazing sealant is used, the fragments near the frame edge cannot readily fall out. If a single glazed light of glass is broken, a windstorm can more easily dislodge some of those pieces if a dry, gasket glazing system is used. If the glass is a double glazed insulating glass unit with both lights of the same thickness and joined at the perimeter with spacer/sealant materials, then there is a high probability that the second light of glass will successfully resist wind load after the first (outboard) light has broken, even though it only has half the strength of the original double glazed insulating glass unit. This can often be seen in photos of hurricane damaged buildings where one light is broken, perhaps from an impact, but the second light has remained intact throughout the storm.

It is important to realize that it takes a very low level of tensile stress, at right angles to the tip of a glass crack, to propagate that crack further. While cracked lights of glass from, say, thermal stress have been seen surviving strong wind loads because the direction of the major bending and membrane stresses in the plate did not happen to be at right angles to the crack tip, it should be recognized that cracked, annealed glass has little or no residual load resistance to tensile stresses across the crack direction. For this reason, annealed glass with a visible crack in it should be removed and replaced as soon as possible.

Heat-strengthened glass breaks in a manner similar to annealed glass when its surface compression is close to the lower end (3500 psi (24 MPa)) of the defined range of 3500 to 7500 psi (24 to 52 MPA) in ASTM C1048. If the surface compression is close to the higher end (7500 psi (52 MPa)) of the defined range in ASTM C1048, then the fracture pattern is more extensive and begins to resemble that of fully tempered glass (10 000 psi (69 MPa) minimum surface compression). With heat-strengthened glass there may be a greater possibility of an island crack pattern occurring (as compared to annealed glass), where a cracked piece may not be directly connected to the frame at the glass edge. Such pieces can readily fall out under light to moderate wind loads, as implied in Fig. 6.9.

Laminated glass with two or more plies of annealed or heat-treated glass, and adhesively connected to the frame at the glass panel perimeter, will usually remain in place under continuing application of the design load, even with all the glass plies broken. Current hurricane load testing and certification procedures can firmly establish such performance levels for particular window designs.

Tempered glass is a 'safety glass' in that it is very unlikely to break, but if it does break it will shatter into a relatively safe mode of 'dice' about the size of cubes of the glass thickness. Under wind load these broken cubes will readily fall out of a single glazed installation, although they often stay in place for some time if they are one light of an insulating glass unit, due to the load resistance of the unbroken adjacent light. The breaking action of FT glass causes multiple very small gaps to appear at each crack. Cumulatively, they have the effect of making the glass expand slightly, in its plane, thus wedging the broken pieces into the frame.

Tempered glass (and heat-strengthened glass to a limited extent, particularly if the surface compressive pre-stress is near the upper limit of

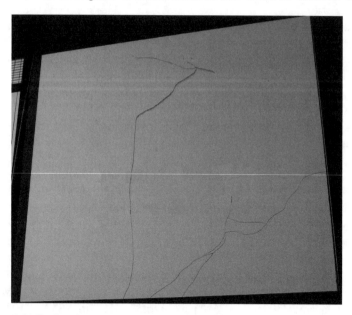

6.9 Broken heat-strengthened glass panel.

7500 psi (52 MPa) for heat-strengthened glass) is susceptible to occasional 'spontaneous breakage' for no obvious reason, i.e. without an obvious overload condition. If circumstances exist that create or develop a very small crack, such as a deep scratch penetrating through the compression layer and just into the tension zone, or edge damage caused by insufficient glazing clearances closing up when concrete beams and floor slabs creep or sag slightly with time, or in a seismic event, then heat-treated glass can suddenly shatter completely. Such sudden, complete breakage is not seen with annealed glass in similar loadings because there are no significant locked-in stresses in the annealed product.

In terms of post-breakage behavior, when both glass plies in a laminated glass unit are broken, the larger particle sizes of heat-strengthened plies will offer better resistance to falling out as a complete 'wet blanket' than when fully tempered glass plies are used. The analysis of post-breakage behavior raises the interesting dilemma where a weaker annealed light is more likely to break than heat-treated glass, but the larger pieces are more likely to remain in a frame after breakage. A stronger heat-strengthened or fully tempered light is less likely to break, but it is more likely to have some or all the pieces fall out once a new load is applied after it has broken. In either case, if the consequences of glass particle fallout are unacceptable, as in overhead glazing or in large viewing windows in an aquarium, then the use of laminated, multi-ply glass is required. With laminated glass the perimeter glazing details and the interlayer material can be selected and tested to be of a type, thickness, and strength to resist loads that may be applied after any or all of the glass plies have broken.

It is very difficult to predict detailed post-breakage behavior of glass. If such details must be known, as, for example, for glazing in hurricane prone areas where the integrity of the building envelope is critical, then laboratory or in situ physical testing of full-scale samples is required. Such test setups must include frames and the glazing sealant connections between the glass and the frame to ensure realistic results.

6.7   Conclusions

A correct glass design will begin with a properly determined wind load (magnitude and gust duration) and will use glass strength charts and factors that recognize the area of glass under stress and the effective duration of the applied load. After considering other factors such as visible light transmission, thermal insulation, solar control, thermal stress, acoustics, aesthetics, electromagnetic frequency transmission, etc., the correct window glass design will also recognize the consequences of any unexpected breakage and determine that they are acceptable and that the final glass type and thickness are adequate.

6.8   Acknowledgement

Figures 6.3 and 6.8 are reprinted, with permission, from ASTM E1300-07e1 *Standard Practice for Determining Load Resistance of Glass in Buildings*, copyright ASTM International, 100 Bar Harbor Drive, West Conshohocken, PA 19428.

6.9   References

1. ASTM E1300, *Standard Practice for Determining Load Resistance of Glass in Buildings*, ASTM International, West Conshohocken, PA, 2007.
2. Beason, W. L., Kohutek, T. L. and Bracci, J. M., Basis for ASTM E1300 glass thickness selection procedure, Civil Engineering Department, Texas A&M University, 1996.
3. Michalske, T. and Bunker, B., The fracturing of glass, *Scientific American*, December 1987.
4. National Standard of Canada CAN/CGSB-12.20-M89 (2), *Structural Design of Glass for Buildings*, The Canadian General Standards Board, 1989.
5. Johar, S., Dynamic fatigue of flat glass, Phase II and III, Final Reports 67039 and 67049, Ontario Research Foundation, Mississauga, Ontario, Canada, 1981 and 1982.
6. ASCE 7-05, *Minimum Design Loads for Buildings and Other Structures*, American Society of Civil Engineers, Reston, VA, 2005.

7.   ASTM E2431-06, *Standard Practice for Determining the Resistance of Single Glazed Annealed Architectural Flat Glass to Thermal Loadings*, ASTM International, West Conshohocken, PA, 2006.
8.   *Glass for Construction*, Libbey-Owens-Ford Company, 1974.
9.   Roark, R. J. and Young, W. C., *Formulas for Stress and Strain*, McGraw-Hill Book Company, New York, 2001.
10. Glass Performance Days Proceedings Conference, Tampere, Finland, organized by Glaston Inc. 2007. Available at: www.glassfiles.com.
11. *ICC Building Codes*, International Code Council Inc., Country Club Hills, IL, 2009.
12. ASTM C1048, *Standard Specification for Heat-Treated Flat Glass*, ASTM International, West Conshohocken, PA, 2004.
13. *Good Glazing Guidelines*, Pilkington, Toledo, OH, 2006. Available at: http://www.pilkington.com/resources/goodglazingguidelinesdoc.pdf
14. *Window Glass Design*, Standards Design Group, 2004. Available at: www.standardsdesign.com.

7

Architectural glass to resist wind-borne debris impacts

D. B. HATTIS, Building Technology Inc., USA

Abstract: The phenomenon of internal pressurization of partially enclosed buildings has been long recognized and understood by engineers and represents more than a threefold increase in internal pressure, which significantly increases the net pressures for which the building envelope must be designed. Unintended internal pressurization of partially enclosed buildings occurs when the building envelope is breached. One type of breach, glass breakage, is most likely to occur at or very close to the time that the building is exposed to its design wind load: breakage during a hurricane or other windstorm as the result of impact from wind-borne debris. In this case the building changes from being enclosed to being partially enclosed and may immediately be exposed to higher internal pressures that may exceed its designed capacity. The first part of this chapter traces the history of standards development and regulation in the United States to address this situation. The second part provides a survey of current design solutions for fenestration to comply with these wind-borne debris impact standards.

Key words: wind-borne debris, impact, pressure cycling, pass/fail criteria.

7.1 Introduction

The phenomenon of internal pressurization of partially enclosed buildings has been long recognized and understood by engineers. When the total area of openings in one wall of a building is over 4 ft^2 and exceeds the total area of openings in the balance of the building envelope by more than 10% internal pressures in the building increase. ASCE 7, *Minimum Design Loads for Buildings and Other Structures* (ASCE, 2005, p. 47), assigns internal pressure coefficients of $+0.18$ and -0.18 for enclosed buildings and $+0.55$ and -0.55, respectively, for partially enclosed buildings. This represents more than a threefold increase in internal pressure, which significantly increases the net pressures for which the building envelope must be designed.

Design professionals who understand this sometimes design partially enclosed buildings to these higher pressures, particularly in those climates where the building envelope can remain partly open all year. One might refer to these situations as voluntarily partially enclosed buildings. However, what of the case where partial enclosure is not voluntary? Unintended partially enclosed buildings occur when the building envelope is breached. There are many ways that components of the building envelope, especially glass, which is a fragile material, can be breached. Some examples are: accidental breakage from a baseball or golf ball, vandalism, a nearby explosion, and, in the case of tempered glass, spontaneous breakage. None of these breaches is likely to occur when the building is subject to its design wind load, and repairs can be made in ample time. One type of glass breakage is most likely to occur at or very close to the time that the building is exposed to its design wind load: breakage during a hurricane or other windstorm as the result of impact from wind-borne debris. In this case the building becomes partially enclosed and may immediately be exposed to higher internal pressures that may exceed its designed capacity.

The first part of this chapter traces the history of standards development and regulation in the United States to address this situation, which one may refer to as nonvoluntary partially enclosed buildings. The second part provides a survey of current design solutions for fenestration to comply with these standards.

7.2   History of wind-borne debris standards development and regulation

7.2.1 Pre-ASTM history of standards

Early attempts by Minor and Hattis at SBCCI (Southern Building Code Congress International)

The earliest test method intended to represent the effects of hurricanes (United States) and severe tropical cyclones (Australia) on the exterior envelope of buildings was developed in Australia in the wake of Cyclone Tracy that hit Darwin in December 1974. Two separate test methods were developed, which subsequently became part of the regulation of construction in Northern Australia. A test method for fenestration consisted of impacting the fenestration specimen with a piece of structural lumber and a test method for roofing systems consisted of an extensive spectrum of positive and negative pressure cycles.

Soon after Cyclone Tracy, Dr Joseph E. Minor, then director of the

Institute for Disaster Research at Texas Tech University, began work on the development of a test method for fenestration performance in hurricanes. This work led to collaboration between Dr Minor and the author of this chapter to propose the test method and related criteria to the Standard Building Code (promulgated by SBCCI). The proposal gained support from some Florida building officials and several members of the Hurricane ad hoc Committee established by SBCCI, most notably from its chair, Charley O'Mealia. The proposal was formally submitted in 1983 and appeared in that year's *SBCCI Blue Book* (proposed code changes). The proposal, which involved impacting glass with a 4100 g (9 lb) 2 in × 4 in lumber, drew immediate opposition from the major glass manufacturer PPG and others, and it was eventually defeated. It is ironic that in August of that year (1983) Hurricane Alicia struck Houston, and caused a dramatic shower of glass from the facades of downtown high rise buildings that was initiated by impacts from gravel roof aggregate from neighboring buildings.

Hurricane Andrew, Miami/Dade, and SBCCI

As reported by the National Oceanic and Atmospheric Administration (NOAA) ten years after the event, Hurricane Andrew:

> ... was the most destructive United States hurricane of record. It blasted its way across south Florida on August 24, 1992. NOAA's National Hurricane Center had a peak gust of 164 mph – measured 130 feet above the ground – while a 177 mph gust was measured at a private home.
> Andrew caused 23 deaths in the United States and three more in the Bahamas. The hurricane caused $26.5 billion in damage in the United States, of which $1 billion occurred in Louisiana and the rest in south Florida. The vast majority of the damage in Florida was due to the winds.

Hurricane Andrew gave birth to the current regulation of building envelope design and construction in hurricane regions of the United States. The process was initiated in three arenas:

- Miami/Dade County, Florida
- SBCCI
- Texas Department of Insurance

At about the same time, the Glass Research and Testing Laboratory (GRTL) of Texas Tech University, under its then director Dr Scott Norville, began a research program that included impacting glazing with 2 × 4 projectiles propelled from an air cannon.

Miami/Dade County enacted its protocol governing the design and construction of building envelopes in December 1993 and it has been in effect ever since. It included impact by a 2 in × 4 in lumber, impact by gravel pebbles (later changed to steel shot), and pressure cycling to a defined pressure spectrum. Broward County adopted a similar standard in early 1994.

SBCCI began developing a standard that became SSTD 12, *SBCCI Test Standard for Determining Impact Resistance from Windborne Debris*, first published in 1994. It included impact by 2 in × 4 in lumber, impact by 2 g steel balls, and pressure cycling to a defined pressure spectrum. By 1997, SSTD 12 included three levels of impact by 2 × 4s as a function of wind speed. An appendix (nonmandatory unless specifically adopted) to the *1999 Standard Building Code* contained requirements that referenced SSTD 12.

In the mid-1990s, the Texas Department of Insurance established design and construction requirements for building envelopes applicable in the barrier islands that referenced SSTD 12.

Moving the development to ASTM

The Miami/Dade protocol was a regulation that applied to a metropolitan jurisdiction located in the most severe hurricane region in the United States, which had just narrowly escaped a catastrophe had Andrew hit it head on. There was political support in Miami/Dade for enacting the regulation and not much vocal opposition. The Texas Department of Insurance regulation was applied to a rather limited area located in a severe hurricane region as well. However, the SBCCI standard and its incorporation into the model code had the potential of impacting a wide geographical area covering many states, including less severe hurricane areas beginning with fastest-mile wind speeds of 90 mph. For this reason, the opposition to regulation of the building envelope was at its highest intensity and most vocal expression at SBCCI. Opposition came from many segments of the construction industry – homebuilders, window manufacturers, glass manufacturers, and others. As a result of apparent deadlock at SBCCI, some proponents and opponents suggested bringing the deliberations into the theoretically more neutral and less political arena of the American Society for Testing and Materials (ASTM).

The issue of establishing a wind-borne debris task group at ASTM was placed on the agenda of Subcommittee E06.51, 'Performance of windows, doors, skylights and curtain wall', at its April 1993 meeting in New Orleans. The meeting was attended by a large group of stakeholders, all of whom had been vocal participants in the often acrimonious debates at SBCCI. Other than general agreement on the need to develop standards for wind-borne debris, there was little agreement on the direction to take and how to proceed.

7.2.2 ASTM standards development history

Pendulum versus air cannon

One of the contentious issues raised in the development of SSTD 12 involved the test apparatus to deliver the impacts. For the large missile impact, Texas Tech's GRTL and some test labs favored the air cannon firing 2 in × 4 in lumber while many window manufacturers and glass manufacturers favored a pendulum. SSTD 12 finally allowed both, based on an equivalency of impact energy (i.e. a heavier pendulum impacting at a lower velocity was deemed to be equivalent to a lighter 2 × 4 impacting at a higher velocity). This issue threatened to block the initiation of work of the ASTM task group, and it was proposed to establish two task groups, with one working on the development of a standard test method using air cannons and other possible 2 × 4 propulsion devices and the other working on development of a standard test method using a pendulum. The writer was voted chair of the former (Task Group E06.51.17) and James Benney of the Primary Glass Manufacturers Council was voted chair of the latter (Task Group E06.51.18). This split enabled each task group to proceed with development of its work plan. Initially both groups intended to develop test methods that represented performance of building elements subjected to wind-borne debris in hurricanes.

Elimination of the pendulum from wind-borne debris simulation

As the work progressed, and as each of the two task groups brought drafts to internal task group ballots and later to Subcommittee E06.51 ballots over the next several years, it became clear that pendulums were not good representations of wind-borne debris in hurricanes. Analysis showed that the velocities of large objects (such as structural debris) carried in a hurricane wind field were on the order of 9–15 m/s (30–50 ft/s), and the velocities of smaller objects (such as gravel) were on the order of 25–30 m/s (80–100 ft/s) or more. The achievement of such velocities by the tip of a pendulum, while theoretically possible, exceeds the limits of safety in most laboratory situations. Thus, the pendulum test method would have to be carried out at lower velocities, and in order to represent wind-borne debris, with higher masses, using either equivalent momentum or equivalent energy to the projectile test method. During the balloting of the pendulum test method, it was argued that there was no basis for assuming that the failure mechanisms of building envelope materials were a function of either momentum or energy. It was concluded that to represent the impacts of wind-borne debris in hurricanes adequately it is necessary to use velocities generally equivalent to the phenomenon itself.

As the result of this process, Task Group E06.51.18 eliminated all references to wind-borne debris and hurricanes from the test method it was developing. It finally became ASTM E2025-99, *Standard Test Method for Evaluating Fenestration Components and Assemblies for Resistance to Impact Energies*. Applications of this test method are currently being discussed in relation to fenestration security.

Separate test method from specification

Task Group E06.51.17 began working soon after the organizational meeting. Early on the decision was made to develop two separate standards – a standard test method (consisting of apparatus and procedure) and a standard specification (consisting of impact and pressure criteria, pass/fail criteria, and product qualification principles). This was a departure from the Miami/Dade protocol and SSTD 12, both of which included all of these elements in a single document. Experience in the development of SSTD 12 and the concurrent efforts to mandate it in the *Standard Building Code* demonstrated that the most contentious subjects were the impact and pressure criteria and the pass/fail criteria. By separating the standards to be developed in this manner the task group felt that the work would be more manageable and that the contentiousness could be contained and overcome.

The task group began working on the standard test method immediately and initiated work on the standard specification in December 1994. It had established as its objective to complete both consensus standards in time for their inclusion as references in the first edition of the ICC *International Building Code* (IBC), scheduled for the year 2000. The decision to separate the two standards proved correct and the target date for completion was achieved.

The standard test method ASTM E1886, *Standard Test Method for Performance of Exterior Windows Curtain Walls, Doors, and Storm Shutters by Missile(s) and Exposed to Cyclic Pressure Differentials*, was first approved in 1997 and the standard specification ASTM E1996, *Standard Specification for Performance of Exterior Windows, Curtain Walls, Doors, and Storm Shutters Impacted by Windborne Debris in Hurricanes*, was first approved in 1999.

7.2.3 ASTM Standard E1886 Test Method

Scope

The scope of ASTM Standard E1886 (ASTM, 1997, p. 1) was defined as follows in Paragraph 1.1 of the standard:

1.1 This test method determines the performance of exterior windows, curtain walls, doors, and storm shutters impacted by missile(s) and subsequently subjected to cyclic static pressure differentials. A missile propulsion device, an air pressure system, and a test chamber are used to model *some conditions which may be representative of windborne debris and pressures in a windstorm environment.* This test method is applicable to the design of entire fenestration or shutter assemblies and their installation. The performance determined by this test method relates to the ability of elements of the building envelope to remain unbreached during a windstorm (*italic emphasis added*).

'Windborne debris' and 'pressures in a windstorm environment' are further elaborated in Paragraphs 5.3 and 5.5 of the standard (ASTM, 1997, pp. 2–3):

> 5.3 The windborne debris generated during a severe windstorm varies greatly, depending upon wind speed, height above the ground, terrain, surrounding structures, and other sources of debris **(4)**. Typical debris in hurricanes consists of missiles including, but not limited to, roof gravel, roof tiles, signage, portions of damaged structures, framing lumber, roofing materials, and sheet metal **(4, 7, 9)**. Median impact velocities for missiles affecting residential structures considered in Ref **(7)** ranged from 9 m/s (30 fps) to 30 m/s (100 fps). The missiles and their associated velocity ranges used in this test method are selected to reasonably represent typical debris produced by windstorms.
>
> 5.5 Cyclic pressure effects on fenestration assemblies after impact by windborne debris are significant **(6–8, 10–12)**. It is appropriate to test the strength of the assembly for a time duration representative of sustained winds and gusts in a windstorm. Gust wind loads are of relatively short duration.

(The boldface numbers in parentheses refer to the list of references at the end of the standard.)

Large missile

ASTM Standard E1886 defined a 'large missile', representative of structural debris in hurricanes, as 'A No. 2 or better Southern yellow pine or Douglas fir 2 × 4 in. lumber'. It did not specify the size, weight, or impact velocity of the large missile, leaving that to the yet to be developed standard specification. Rather, it specified a range of sizes, weights, and velocities, based on experience and discussions to date at Miami/Dade and SBCCI, as follows:

Size:
    1.2 m (4.0 ft) ≤ length ≤ 4.0 m (13.2 ft)
Weight:
    2050 g (4.5 lb) ≤ mass ≤ 6800 g (15.0 lb)
Velocity:
    0.1 ≤ impact speed ≤ 0.55 times the basic 3 s gust wind speed

Note that the 2050 g (4.5 lb) 2 × 4 was the smallest specified in SSTD 12 and that the 6800 g (15.0 lb) 2 × 4 had been specified by consultants for hospitals in south Florida (even though not mandated by Miami/Dade).

Small missile

ASTM Standard E1886 defined a 'small missile', representative of gravel roof ballast in hurricanes, as a 'solid steel ball having a mass of 2 g (0.004 lb) with an 8 mm (5/16 in.) nominal diameter'. The impact velocities were defined as a range, as they were for the large missile:

Velocity:
    0.4 ≤ impact speed ≤ 0.85 times the basic 3 s gust wind speed

Other missiles

The task group discussed other missiles such as those that might represent roof tiles or shingles more appropriately than a 2050 g (4.5 lb) 2 × 4. Being unable to agree on a standard missile of this type, the following sentence was placed in Paragraph 6.2.7.3 of E1886 (ASTM, 1997, p. 3):

6.2.7.3 *Other* Missile—Any other representative missile with mass, size, shape, and impact speed as a function of basic wind speed determined by engineering analysis such as Ref **(9)**.

Missile propulsion device

ASTM Standard E1886 defines the missile propulsion device in performance terms as 'any device capable of propelling the missile at a specified speed, orientation, and impact location'. It also specifies that the missile should not be accelerating upon impact due to gravity, which rules out a drop test apparatus. As stated above, it was assumed that a pendulum could not be used to achieve the range of specified missile speeds safely.

Appendix X1 of ASTM Standard E1886 provides some examples of missile propulsion devices that already had been developed. For the large missile, two examples are provided: air cannon and bungee test apparatus. For the small missile, an air cannon apparatus is provided as an example.

Table 7.1 ASTM Standard E1886, pressure cycling spectrum (extracted with permission from Table 2, E1996.08, copyright ASTM International, 100 Barr Harbor Drive, West Conshohocken, PA 19428, USA)

Loading Sequence	Loading Direction	Air Pressure Cycles	Number of Air Pressure Cycles
1	Positive	0.2 to 0.5 P_{pos}	3500
2	Positive	0.0 to 0.6 P_{pos}	300
3	Positive	0.5 to 0.8 P_{pos}	600
4	Positive	0.3 to 1.0 P_{pos}	100
5	Negative	0.3 to 1.0 P_{neg}	50
6	Negative	0.5 to 0.8 P_{neg}	1050
7	Negative	0.0 to 0.6 P_{neg}	50
8	Negative	0.2 to 0.5 P_{neg}	3350

Pressure cycling

As stated above, the ASTM Standard E1886 test method follows the specified missile impact with a pressure cycling regime designed to represent wind gusts in a hurricane. The cyclic pressure spectrum specified in E1886 is the same as that specified in Miami/Dade and in SSTD 12. It is based on research reported by Letchford and Norville (1994), 'Wind pressure loading cycles for glazing during hurricanes'. It consists of 9000 pressure cycles starting with increasing positive pressure cycles and followed by decreasing negative pressure cycles. Pressure cycle duration is defined as no less than 1 s nor more than 5 s. Pressure cycles are specified as a sequence of pairs of pressures that are defined as a fraction of *P*, where *P* is the design pressure specified in ASCE 7 or the local code for the assembly being tested. Table 7.1 shows the pressure cycling spectrum specified in ASTM Standard E1886.

Subsequent modifications

Following its initial publication in 1997, ASTM Standard E1886 was modified twice, and the current standard carries the date 2004. Two minor changes were made in 2002. The first was to achieve consistency with ASTM Standard E1996, and consisted of a minor adjustment to the range of velocities specified for the small missile. The second was to accommodate labs that could not reverse pressure in their testing apparatus, and permitted the removal, reversal, and reinstallation of the test specimen in the test chamber between the positive and negative pressure cycles. In 2004 the term 'storm shutters' was changed to 'impact protective systems'.

7.2.4 ASTM Standard E1996 Specification

Introduction

Task Group E06.51.17 began formal deliberations on the standard specification in 1994. These deliberations benefited from the group's collaborative work on the test method as well as criteria used in Miami/ Dade and proposed for the *Standard Building Code*. The task group was joined by Lawrence Twisdale of Applied Research Associates, who provided an analytical basis that supported the judgments made by the task group. He had co-authored a report in 1996, 'Analysis of hurricane windborne debris impact risk for residential structures' (Twisdale *et al.*, 1996), which provided support for parts of the standard, as discussed in Appendix X2 of E1996 and as quoted in Section 7.2.7 below.

ASTM E1996 was first published in 1999. Since then there have been several revisions, with the most recent edition dated 2008. The following discussion presents the current edition of the standard, with mention of earlier editions where appropriate.

Impact parameters

The task group considered which of the theoretical parameters of traditional wind design, as defined and specified in ASCE 7, should be included in the specification of debris impact. The first was 'basic wind speed', which was included because it was clear that wind-borne debris velocity in a windstorm is directly related to wind speed, and SBCCI had already related specified impact mass and impact energy to basic wind speed. ASTM Standard E1996 initially defined three wind zones, which were similar to those in SSTD 12, as follows:

- Wind Zone 1: 110 mi/h ≤ basic wind speed < 120 mi/h and Hawaii
- Wind Zone 2: 120 mi/h ≤ basic wind speed < 130 mi/h at greater than 1.6 km (1 mile) from the coastline
- Wind Zone 3: basic wind speed ≥ 130 mi/h or basic wind speed ≥ 120 mi/ h and within 1.6 km (1 mile) of the coastline

In 2002 Wind Zone 3 was divided into two, Wind Zones 3 and 4, as the result of discussions with Miami/Dade staff that were initiated in 2001 (to explore the changes required to enable Miami/Dade to reference the ASTM standards and discontinue their own protocol), as follows:

- Wind Zone 3: 130 mi/h (58 m/s) ≤ basic wind speed ≤ 140 mi/h (63 m/s) or 120 mi/h (54 m/s) ≤ basic wind speed ≤140 mi/h (63 m/s) and within 1.6 km (1 mile) of the coastline
- Wind Zone 4: basic wind speed > 140 mi/h (63 m/s)

The second parameter considered for inclusion in ASTM E1996 was 'exposure category', which was excluded as an impact parameter following internal task group discussions because its effect on wind-borne debris was considered complex and indirect. On the one hand, wind speed is greater in more open exposures, but, on the other hand, rougher more built-up exposures are more likely to produce larger quantities of wind-borne debris in a windstorm.

The third parameter considered for inclusion was 'elevation above the ground', which was included because wind-borne debris is likely to be more abundant and more massive closer to the ground, and both Miami/Dade and SBCCI had related impacts to elevation above the ground. ASTM Standard E1996, like SSTD 12, simplified the elevation parameter by defining two elevations:

- $\leq$ 9.1 m (30 ft) above the adjacent ground level, where large missile impacts are specified
- $>$ 9.1 m (30 ft) above the adjacent ground level, where small missile impacts are specified

The fourth parameter considered for inclusion was 'occupancy classification', which was included in ASTM 1996 because the task group agreed that in addition to the *basic protection* specified for most buildings, certain categories of buildings required *enhanced protection* from wind-borne debris and others required *no protection* (neither basic nor enhanced). Neither Miami/Dade nor SSTD 12 included this parameter. Enhanced protection was defined in ASTM Standard E1996 (ASTM, 1999, p. 3) as follows:

Enhanced protection (essential facilities)—Buildings and other structures designated as essential facilities, including, but not limited to, hospital and other health care facilities having surgery or emergency treatment facilities; fire, rescue and police stations and emergency vehicle garages; designated emergency shelters; communications centers and other facilities required for emergency response; power generating stations and other public utility facilities required in an emergency; and buildings and other structures having critical national defense functions.

In 2003 slight modifications were made to this definition in order to increase consistency with ASCE 7. Jails and detention facilities were added, and 'other health facilities having surgery' were deleted.

No protection was specified in ASTM Standard E1996 (ASTM, 1999, p. 3) as follows:

Unprotected—Buildings and other structures that represent a low hazard to human life in a windstorm including, but not limited to: agricultural facilities, production greenhouses, certain temporary facilities, and storage facilities.

Impact levels

ASTM Standard E1996 initially defined three large missiles impact levels (B, C, and D), all within the range of large missiles specified in ASTM Standard E1886 and discussed above, and one small missile (A). The four missiles initially applied to both fenestration and skylights. In 2001 two changes were made related to impact levels. The first consisted of a reduction of specified large missile impacts for rooftop skylights in one- and two-family dwellings. The reduction was supported by reference to the Twisdale *et al.* 1996 report, based on reduced risk of failure because residential rooftop skylights are exposed to fewer missile sources by virtue of elevation and because residential rooftop skylights constitute a smaller percentage of overall roof surface than windows do as a percentage of overall wall surface. In order to accomplish the reduced impact in Wind Zone 1 (lowest) the second related change was made. It was necessary to use a smaller 2 × 4 missile and a 910 g (2.0 lb) missile was specified. (The new missile was designated B and the original missiles B, C, and D were redesignated C, D, and E, respectively, which, unintentionally, has caused a great deal of confusion.) The five missiles and their designation are shown in Table 7.2 (ASTM, 2008, p. 4). The assignment of each of these missiles to test specimens, as a function of wind zone, elevation above the ground, and level of protection, as defined above, is shown in Tables 7.3 and 7.4 (ASTM, 2008, p. 4), the latter for residential skylights.

There was extensive discussion within the task group that led to the establishment of these levels, and there was continuous opposition primarily from the homebuilders. The impact levels are not very different from those used in SSTD 12, and the Twisdale *et al.* 1996 report mentioned above provided the basis for confirming that these levels were within the range of reasonableness (i.e. 'in the ballpark'). This is discussed in Appendix X2 of ASTM Standard E1996, 'Impact risk analysis', which is reproduced in full in Section 7.2.7 of this chapter.

Number of impacts

Initially, ASTM E1996 specified one impact on each of three identical specimens – one center impact and two opposite corner impacts – for the large missile and three impact locations for each of three identical specimens for the small missile. In 2002, as the result of discussions with Miami/Dade staff mentioned above, the standard was modified for Wind Zone 4 to

Table 7.2 ASTM Standard E1996-01, missiles (extracted with permission from Table 2, E1996-08, copyright ASTM International, 100 Barr Harbor Drive, West Conshohocken, PA 19428, USA)

Missile Level	Missile	Impact Speed (m/s)
A	2 g ± 5 % steel ball	39.62 (130 f/s)
B	910 g ± 100 g (2.0 lb ± 0.25 lb) 2 × 4 in. 52.5 cm ± 100 mm (1 ft − 9 in. ± 4 in.) lumber	15.25 (50 f/s)
C	2050 g ± 100 g (4.5 lb ± 0.25 lb) 2 × 4 in. 1.2 m ± 100 mm (4 ft ± 4 in.) lumber	12.19 (40 f/s)
D	4100 g ± 100 g (9.0 lb ± 0.25 lb) 2 × 4 in. 2.4 m ± 100 mm (8 ft ± 4 in.) lumber	15.25 (50 f/s)
E	4100 g ± 100 g (9.0 lb ± 0.25 lb) 2 × 4 in. 2.4 m ± 100 mm (8 ft ± 4 in.) lumber	24.38 (80 f/s)

Table 7.3 Missile levels and where they apply (extracted with permission from Table 3, E1996.08, copyright ASTM International, 100 Barr Harbor Drive, West Conshohocken, PA 19428, USA)

Level of Protection	Enhanced Protection (Essential Facilities)		Basic Protection		Unprotected	
Assembly elevation	≤(30 ft) 9.1 m	>(30 ft) 9.1 m	≤(30 ft) 9.1 m	>(30 ft) 9.1 m	≤(30 ft) 9.1 m	>(30 ft) 9.1 m
Wind Zone 1	D	D	C	A	None	None
Wind Zone 2	D	D	C	A	None	None
Wind Zone 3	E	D	D	A	None	None
Wind Zone 4	E	D	D	A	None	None

Table 7.4 Missile levels for rooftop skylights in one- and two-family dwellings (extracted with permission from Table 4, E1996-08, copyright ASTM International, 100 Barr Harbor Drive, West Conshohocken, PA 19428, USA)

Level of Protection	Basic Protection	
Assembly elevation	≤(30 ft) 9.1 m	>(30 ft) 9.1 m
Wind Zone 1	A	A
Wind Zone 2	B	A
Wind Zone 3	C	A
Wind Zone 4	D	A

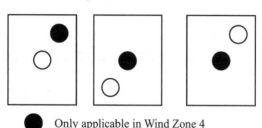

● Only applicable in Wind Zone 4

7.1 Impact locations for large missile test (extracted with permission from Figure 1, E1996-08, copyright ASTM International, 100 Barr Harbor Drive, West Conshohocken, PA 19428, USA).

require two impacts on each of three identical specimens (see Fig. 7.1) as well as additional impacts specified on mullions.

Pass/fail criteria and product qualification

Three test specimens must be submitted for each of the large and small missile impact tests, to be followed by the pressure cycling. All three specimens must pass the test for acceptance of the product. (This was later changed to three out of four specimens.) The pass/fail criteria in ASTM Standard E1996 were as follows:

1. For fenestration assemblies and nonporous shutter assemblies: no tear longer than 130 mm (5 in) or no opening through which a 76 mm (3 in) sphere can freely pass following the impact and pressure cycling.
2. For porous shutter assemblies tested independently of the fenestration they are protecting: no penetration of the innermost plane during the impact test(s) and no horizontally projected opening through which a 76 mm (3 in) sphere can freely pass following the impact and pressure cycling.

In 2002, as the result of discussions with Miami/Dade staff mentioned above, an additional criterion was established for Wind Zone 4 – no penetration of the innermost plane of the specimen was allowed.

Initially, in addition to qualifying products represented by the three specimens to successfully pass the test, the following products are also qualified:

1. Fenestration assemblies with thicker or equal glazing and thicker or equal interlayer of the same glass type and treatment, provided the glazing detail is unchanged.
2. Fenestration assemblies of the same type that contain smaller sashes, panels, or lites at equal or lower design pressures, provided the same

methods of fabrication are used and the anchorage of the lites is unchanged.

3. Fenestration assemblies with the same glazing type and treatment that are tinted, heat absorbing, reflective, or otherwise aesthetically modified.

4. Successful tests of a fenestration assembly that contains construction to improve thermal efficiency of frame or sash qualify other assemblies without such construction.

5. Shutter assemblies of the same or less area and the same or greater section modulus, provided the construction details and reinforcement are unchanged.

In 2008 Annex A, 'Fenestration substitutions' (ASTM, 2008, pp. 6–10) was added to replace and expand the qualification criteria for fenestration products included in the original standard. The purpose was to rationalize the process of acceptance of changes made to approved products, and eliminates the need, as much as possible, to retest three specimens each time a change is made.

7.2.5 References to ASTM standards in other standards and regulations

IBC and IRC

As stated previously, Task Group E06.51.17 had established the objective of completing the standard test method and standard specification, ASTM Standards E1886 and E1996, in time for their inclusion by reference in the first editions of the *International Building Code* (IBC) and the *International Residential Code* (IRC). This objective was achieved. Both the IBC 2000 and the IRC 2000 contained references to ASTM Standard E1886-97 and ASTM Standard E1996-99.

In the 2003 editions of the IBC and IRC, the reference to E1996 was updated to the 2001 edition, but was still not up to date. Section 1609.1.4 of the IBC 2003 establishes requirements for the protection of openings in 'windborne debris regions', which are defined as 'areas within hurricane-prone regions within 1 mile of the coastal mean high water line where the basic wind speed is 110 mi/h or greater, or where the basic wind speed is equal to or greater than 120 mi/h, or Hawaii'. It requires impact protection for glazing in the lower 60 ft of buildings, but allows designing the buildings for internal pressurization (i.e. assuming the glazing is an opening) as an alternative to impact protection. Section 1609.1.4 specifies that the impact protection meets the requirements of the ASTM standards or 'of an approved impact-resisting standard', the latter presumably allowing the use of the Miami/Dade protocol or SSTD 12. Section R301.2.1.2 of the IRC 2003 is similar to Section 1609.1.4 of the IBC, except that no alternative to the ASTM standards is mentioned.

In the 2006 edition of the IBC the references to E1886 and E1996 were updated to the 2004 editions, which is up to date for the former but not for the latter. Section 1609.1.4 was revised to Section 1609.2. The section eliminates the alternative of designing the buildings for internal pressurization, and adds the requirement for impact protection for glazing above 60 ft if it is located no more than 30 ft above aggregate surface roofs located within 1500 ft of the building.

ASCE 7

ASCE 7, *Minimum Design Loads for Buildings and Other Structures*, began addressing the issue of impact from wind-borne debris in its 1995 edition. ASCE 7-95 includes a discussion in the Commentary of the vulnerability of glazing to impact from wind-borne debris in the lower 60 ft of a building, and the resultant potential for internal pressurization.

ASCE 7-98 contains a definition of 'windborne debris regions' that is identical in substance to and was undoubtedly the source of the definition used in the I-codes. Paragraph 6.5.12.2.1 (ASCE, 1998) states that for '... buildings sited in windborne debris regions, glazing in the lower 60 ft (18.3 m) that is not impact resistant or protected with an impact resistant covering ... shall be treated as an opening ...' for purposes of potential increase in internal pressurization. This subject is further discussed in Commentary C6.5.11.1. 'Impact resistant glazing' and 'impact resistant covering' are defined in terms of '... an approved test method to withstand the impact of windborne missiles likely to be generated in windborne debris regions during design winds'. ASCE 7-98 contains no specific reference to ASTM or any other standards in this regard.

During this period, collaboration between ASTM Task Group E06.51.17 and the ASCE wind committee was initiated. It was agreed that ASTM standards should continue to define the impact and pressure cycling test method as well as specify impact levels and pass/fail criteria, and that ASCE 7 would define where to require wind-borne debris protection (i.e. the scoping). As a result, ASCE 7-02 (ASCE, 2002) is much more specific on the subject than prior editions. Section 6.7 of ASCE 7-02 includes references to ASTM Standards E1886-97 and E1996-99. (Note that these references were out of date at the time of publication of ASCE 7-02.) Paragraph 6.5.9.3 is entitled 'Windborne debris', and it requires impact protection of glazed openings in Categories II, III, and IV buildings, including those glazed openings over 60 ft above the ground if they are located up to 30 ft above 'aggregate surface roof debris located within 1,500 ft of the building'. In

Category II and III buildings (i.e. nonessential facilities) impact protection need not be provided if the glazing is assumed to be an opening for purposes of internal pressurization. Impact protection is required in terms of reference to the two ASTM standards 'or other approved test methods and performance criteria'. However, Note 2 to the paragraph appears to limit the applicability of 'other approved performance criteria' by stating that 'levels of impact resistance shall be a function of missile levels and wind zones specified in' ASTM Standard E1996-99.

The Commentary to ASCE 7-02, Section C6.5.9, expands further on the subject by including the definition of wind zones from ASTM Standard E1996-99 and reproducing adaptations of the E1996 tables of impact levels and missile levels.

In ASCE 7-05, Paragraph 6.5.9.3 (ASCE, 2005, p. 27) is revised with the effect of eliminating the option in Category II and III buildings of designing for internal pressurization in lieu of impact protection. While the references to ASTM E1886 and E1996 are updated to 2002 and 2003, respectively, they are still not up to date.

Changes needed

It is clear from the preceding discussion that developments of the two ASTM standards, the I-codes (i.e. IBC and IRC), and ASCE 7 have not been well coordinated to date. The references to the ASTM standards are out of date, although two important early revisions to E1996 – the residential skylight provisions and the Wind Zone 4 provisions – are required by IBC 2006 and by ASCE 7-05. The scoping requirements in the IBC and ASCE 7 were inconsistent in 2003 as applicable over 60 ft above the ground. However, they have been made consistent in 2006. It is also unclear as to what alternative standards to ASTM are acceptable under either the IBC or ASCE 7.

Florida Building Code

The state with the largest percentage of its area subject to hurricanes in the United States is Florida. As stated earlier, Miami/Dade County has been enforcing its own wind-borne debris protocol since 1993. Therefore it is useful to include here a brief discussion of the *Florida Building Code*. Florida adopted a statewide preemptive building code in 2001. Subsequent editions of the code are 2004 and 2007. The *Florida Building Code* is based on the IBC with some amendments. As regards wind-borne debris requirements, one of the amendments to the *Florida Building Code* is the definition of a high-velocity hurricane zone (HVHZ), which consists of Miami/Dade and Broward Counties, within Wind Zones 3 and 4. Miami/Dade's protocol is applicable within the HVHZ. The rest of the state follows the requirements of the IBC, and the ASTM standards referenced therein.

It should be noted that Wind Zone 4 requirements were introduced into E1996 in response to requests from Miami/Dade and in an attempt to permit Miami/Dade to substitute E1996 for it own protocol. However, not all of Miami/Dade's requests were accepted into the ASTM standard. As a result, there are some significant differences between the HVHZ and the Wind Zone 4 requirements in the *Florida Building Code*. These mainly relate to the operability of fenestration following the impact and pressure cycling test, to the disengagement of fasteners during the testing, and to the applicability of the requirements to solid elements of the envelope such as EIFS and vinyl siding.

Some of the Miami/Dade requirements that are currently included in the Wind Zone 4 requirements, such as offset requirements for impact protective systems (hurricane shutters), have proven to be burdensome to product manufacturers within Wind Zone 4 but outside the HVHZ. This has led to a current proposed change to E1996 that, if adopted, would allow some of the *Florida Building Code* HVHZ requirements to be 'optional additional pass/fail criteria' within Wind Zone 4.

7.2.6 Currently balloted changes to ASTM standards

Two changes to ASTM Standard E1996 are currently being balloted. The ballots will be adjudicated at the April 2009 meetings of ASTM Committee E06. While these changes cannot be discussed in detail at this time, one of them is the attempt to separate the Florida HVHZ requirements from the general requirements of Wind Zone 4, while the other is a new Annex on shutter substitutions to complement the Annex on fenestration substitutions added in 2008.

7.2.7 ASTM E1996 Appendix on impact risk analysis

The following section is extracted, with permission, from E1996-99 (ASTM, 1999), copyright ASTM International, 100 Barr Harbor Drive, West Conshohocken, PA 19428.

X2. IMPACT RISK ANALYSIS

X2.1 *Summary of Risk Parameters* in Ref **(5)**–The report discusses the following parameters that affect the risk of building damage from windborne debris:

X2.1.1 Wind velocity,

X2.1.2 Type and quantity of missiles in the wind-field generated from ground sources,

X2.1.3 Type and quantity of missiles in the wind-field generated from building sources, as function of the quality of construction,

X2.1.4 Density of buildings,

X2.1.5 Shape and height of buildings, and

X2.1.6 Percentage of glazed openings.

X2.2 The report combines a hurricane wind field model, a missile generation model, a missile trajectory model and an impact model to produce a risk analysis. The output is expressed in terms of curves of specified impact energy resistance or impact momentum resistance levels plotted on a graph with reliability (R) (from 0.75 to 1.00) on the vertical axis and wind velocity (from 110 to 170 mph peak gusts) on the horizontal axis. Plots have been generated for single story detached residential buildings, for two different values for the quality of construction and density of buildings, and three different values for percentage of glazed openings.

X2.3 *The Performance Objective of This Specification*

X2.3.1 This specification establishes missile impact criteria for all building types and occupancies. The antecedents for this effort are the criteria established in Australian National Standards **(6)**, the Florida counties of Dade **(1)** and Broward **(2)**, in SBCCI Standard SSTD 12 **(3)**, and in the Texas Department of Insurance Building Code for Windstorm Resistant Construction **(4)**. All of these are based on analysis and judgment of experts after many years of windstorm study. The Twisdale *et al.* study represents new inputs into this body of analysis and experience. Since it so far has covered only a very limited range of buildings out of the total scope of this specification, its application to the development of this specification has also required a degree of judgment.

X2.3.2 The energy and momentum curves included in the Twisdale *et al.* **(5)** report are referenced to a zero energy or momentum curve, that can be interpreted as the reliability achieved at various wind speeds when no impact resistance is provided. Other curves describe reliability versus wind speed at increasing amounts of impact resistance, for example 10, 20, 50, 100, 200 and 300 lb of momentum. All the curves illustrated by Twisdale *et al.* **(5)** including the zero resistance curve, demonstrate reliability above 0.85 at 110 mph wind speed. Reliability diminishes rapidly, with varying slopes, at higher wind speeds.

X2.3.3 Two approaches can be taken to using these curves to inform the specification process: the absolute reliability approach and the relative improvement approach.

X2.3.4 *The absolute reliability approach* establishes the objective of achieving a specified level of reliability, say 0.90, by specifying the appropriate impact resistance for different wind speeds, and, possibly, building types. This approach is attractive because it enables the definition of reliability to be consistent with the reliability objective of traditional structural design. However, it has two disadvantages in this case:

X2.3.4.1 The curves plotted are actually average values and should be thought of as

broad fuzzy bands with large confidence bounds due to the many uncertainties embedded in the analytical models that produce them. Therefore, establishing a specified reliability level may be misleading without extensive qualifying statements.
X2.3.4.2 The curves diminish so fast at higher wind speeds that the levels of resistance required to achieve high values of reliability at these wind speeds would require impact energies and momenta far in excess of anything considered heretofore, and possibly in excess of the capabilities of the apparatus specified in Test Method E1886.
X2.3.5 *The relative improvement approach* takes its cue from the zero protection curves, and establishes the objective of achieving a specified proportional improvement in reliability. A 50% improvement, .50 to .75 (sic), 0.60 to 0.80, 0.70 to 0.85, 0.80 to 0.90 etc., could be the objective. The curves illustrated by Twisdale *et al.*, for the limited range of parameters analyzed, suggest that a 50% or better improvement can be achieved by providing impact protection from a 4100 g (9 lb) 2 by 4 traveling at 15.24 m/s (50 ft/s). This is of the same order of magnitude included in the Australian, SBCCI, Florida, and Texas standards.
X2.3.6 Thus, the proposed specification can be justified on the basis of the relative improvement approach and its relation to previous research and antecedents. It can be further refined as more analytical information is developed.
. . .

References

(1) "Section 2315 Impact Tests for Windborne Debris", South Florida Building Code – Dade County Edition, Metro Dade County, Miami, FL, 1994, pp. 23-33–23-38.
(2) "Section 2315 Impact Tests for Windborne Debris and Section 2316 Impact Test Procedures," *South Florida Building Code* – Broward County Edition, Broward County Board of Rules and Appeals, Ft Lauderdale, FL, 1994, pp. 23-24–23-21.
(3) *SBCCI Test Standard for Determining Impact Resistance from Windborne Debris,* Southern Building Code Congress International, Inc., 900 Montclair Road, Birmingham, AL 35213-1206, 1994.
(4) *Building Code for Windstorm Resistant Construction,* Texas Department of Insurance, 33 Guadalupe Street, Austin, TX 78714-9104, 1997.
(5) Twisdale, L.A., Vickery, P.J., and Steckley, A.C., *Analysis of Hurricane Windborne Debris Impact Risk for Residential Structures,* Applied Research Associates, Inc., Raleigh, NC, March 1996.
(6) Standard Australia, *Australian Standard SAA Loading Code, Part 2: Wind Loads,* North Sydney, New South Wales, Australia 2060, 1989.

7.2.8 Conclusions

In this section of the chapter the writer traces the history of the development of standards and code requirements for the design of glazed openings in hurricane regions in the United States. Glazed openings must be designed to perform successfully when subjected to a specified test method. Requirements are described in ASTM standards E1886 and E1996. Recent editions of the building codes and other applicable standards have referenced these standards. However, the references are not up to date with the current editions of the ASTM standards, and in some instances the various documents need better coordination on the subject. While the building codes and the standards referenced therein are to be considered as the applicable requirement, the practitioner reader should refer to the current editions of the ASTM standards in discussions with code enforcement officials.

7.3   Survey of current design solutions

7.3.1 Wind zones, their geographic locations, and respective requirements

As discussed above, codes and standards in the United States have defined four wind zones for the purpose of regulating resistance to debris impact in hurricanes.

Wind Zone 1 includes one mile from the coastline of the Atlantic coast from Massachusetts to Virginia, and additionally includes Hawaii. (Note that this definition is in the ICC codes and in ASCE 7, and that ASTM E1996 extends Wind Zone 1 further inland than one mile from the coastline.)

Wind Zone 2 includes the Atlantic coast from North Carolina to central Florida, parts of eastern Long Island, parts of south central Florida and the Gulf coast from Florida to the Mexican border.

Wind Zone 3 includes the eastern tip of Long Island one mile from the coastline, parts of the Atlantic coast in North Carolina, South Carolina, and northeast Florida, South Florida, parts of the west coast and pan handle of Florida, and parts of the Texas coastline.

Wind Zone 4 includes the southeast and southern tip of Florida and some barrier islands in North Carolina and elsewhere. Within Wind Zone 4, the High Velocity Hurricane Zone consists of Miami/Dade and Broward counties.

The small missile impact requirements are identical in all four wind zones. The large missile impact requirements are identical in Wind Zones 1 and 2, and the only difference in their respective requirements is the higher pressures in the pressure cycling test in Wind Zone 2. The large missiles in Wind Zones 3 and 4 are identical and larger than those in Wind Zones 1 and 2, but two impacts per specimen are required in Wind Zone 4. Higher pressures are required in the Wind Zone 4 pressure cycling test. The pass/fail criteria are stricter in Wind Zone 4 than in the lower wind zones, and stricter yet in the High Velocity Hurricane Zone within Wind Zone 4.

7.3.2 Current design solutions by wind zone

There is no generally available information on design differentiation by wind zones for both commercial and residential applications. Product literature is unclear on the subject, and extensive market research would be required to develop product taxonomy by wind zone. In fact, it is likely that the hurricane impact resistance market has not matured to the point where such taxonomy exists. The following discussion is based on limited anecdotal evidence.

Commercial fenestration and curtain wall

The author examined recent commercial specifications for hotel projects in Naples, Florida (Wind Zone 3), and New Orleans (Wind Zone 2 or 3). In both cases documentation certifying the fenestration was compliance with Miami/Dade (Wind Zone 4, HVHZ requirements). The certifying test laboratory reported that about 95% of the commercial products are tested for Wind Zone 4 requirements. This suggests, pending further market research, that commercial product manufacturers market Wind Zone 4 products in all wind zones. Typically the frame manufacturers will design the metal to meet their highest desired performance target, with the largest opening sizes they anticipate providing for each system. The joints, fasteners, and anchoring are calculated for optimal performance and efficiency. The glass is designed keeping in mind size, loads, and available glass bite in compliance with ASTM E1300. Minimum or maximum glass sizes are not required for testing. A manufacturer will test the largest size for a configuration as all smaller sizes of the same configuration will be automatically qualified. Various laminate configurations are normally tested to allow the greatest design flexibility for those utilizing the framing system.

The missile impact resistance is generally provided by laminated glass of 7/16–9/16 inch thickness. Glass types in the plies of the laminate construction vary depending on the design load. Annealed glass is not typically used for commercial applications because of the lower glass strength and the long sharp shards into which it breaks. Fully tempered glass as plies of the laminated glass are also avoided due to the small break pattern of the glass and can contribute to the flexibility of the laminate once broken. Therefore, heat-strengthened glass is the most common glass used for the plies of laminated glass as it gives the best tradeoff for glass strength and a favorable break pattern. Interlayer thickness of 0.060 in (1.52 mm) or greater (0.060 in or less for some products for a small missile, 0.075–0.090 in or greater for higher impacts for a large missile) is specified for missile impact. The actual thickness and type of the interlayer is selected based on the missile size and speed, glass size, design pressure, frame system, available glass bite, and type of glazing (wet or dry). Where required by code to provide added thermal insulations, a sealed insulating glass unit is of $1\frac{5}{16}$ inch overall thickness, generally composed of $\frac{1}{4}$ inch outboard glass (heat-strengthened, fully tempered, or laminated), $\frac{1}{2}$ inch space, and the appropriately sized (typically $\frac{9}{16}$ inch) laminated glass inboard. Laminated lites are most commonly placed inboard to reduce the amount of glass fragmentation that is projected to the interior of the facility should impact occur. Occasionally, facilities will require design for the reduction of glass fallout from a building should a storm with damaging debris occur. Laminate positioning (outboard versus inboard) or laminated lites on both sides of the insulating space are options that are utilized in order to meet this design criteria. In this case, the second laminated lite is typically laminated with minimal interlayer and glass thickness to ensure shard retention only.

Residential fenestration

The major residential window manufacturers in the United States all offer impact resistant windows in hurricane regions. One manufacturer has published a handbook containing multiple window configurations and sizes. The handbook classifies the windows into three impact zones, 2–4, which the author assumes to correspond to Wind Zones 2, 3, and 4. Other than the higher design wind pressures indicated respectively for the higher impact zones, there is no clear indication how the products differ by wind zone. This is due to changes required to meet the structural requirements beyond debris impact. Where required by the code to provide added thermal insulations, sealed insulating glass is used. As the impact resistant window design continues to meet energy code requirements the differentiation of products will evolve. The need to minimize cost and drive efficiency in products will mandate product differentiation based on wind zone. Further market research is needed to identify this evolving differentiation.

7.4   References

ASCE (1998) ASCE 7-98, *Minimum Design Loads for Buildings and Other Structures*, American Society of Civil Engineers, Reston, VA.

ASCE (2002) ASCE 7-02, *Minimum Design Loads for Buildings and Other Structures*, American Society of Civil Engineers, Reston, VA.

ASCE (2005) ASCE 7-05, *Minimum Design Loads for Buildings and Other Structures*, American Society of Civil Engineers, Reston, VA.

ASTM (1997) ASTM E1886-97, *Standard Test Method for Performance of Exterior Windows, Curtain Walls, Door and Storm Shutter Impacted by Missile(s) and*

Exposed to Cyclic Pressure Differentials, American Society for Testing and Materials, West Conshohocken, PA.

ASTM (1999) ASTM E1996-99, *Standard Specification for Performance of Exterior Windows, Curtain Walls, Doors and Storm Shutters Impacted by Windborne Debris in Hurricanes*, American Society for Testing and Materials, West Conshohocken, PA.

ASTM (2008) ASTM E1996-08[e2], *Standard Specification for Performance of Exterior Windows, Curtain Walls, Doors, and Impact Protective Systems Impacted by Windborne Debris in Hurricanes*, American Society for Testing and Materials, West Conshohocken, PA.

Letchford, C. W. and Norville, H. S. (1994) Wind pressure loading cycles for glazing during hurricanes, *Journal of Wind Engineering and Industrial Aerodynamics*, 53, 189–206.

Twisdale, L A., Vickery, P. J. and Steckley, A. C. (1996) Analysis of hurricane wind-borne debris for residential structures, Report prepared for Applied Research Associates, Inc., Raleigh, NC.

8

Glazing systems to resist windstorms on special buildings

J. E. MINOR, Rockport, USA

Abstract: Building envelopes designed and installed in accordance with model building code provisions are the minimum acceptable solution permitted by law. Buildings with special functions or special importance are commonly designed to a significantly higher level of windstorm resistance. This chapter outlines a rational approach to designing architectural glazing systems for building envelope systems that 'transcend the code' by meeting site-specific design requirements formulated to preserve building function during extreme windstorms.

Key words: glazing, windstorms, window glass, wind risk, glazing systems, buildings.

8.1   Introduction

Commonly, building envelopes are designed in accordance with building code provisions that represent the minimum acceptable standard permitted by applicable law or ordinance. A new standard of practice has evolved wherein buildings of special importance or buildings that will house special functions are designed to higher levels of resistance than are required by these minimum standards. With respect to the effects of wind, this new standard of practice requires a site-specific *wind analysis*, the selection of an appropriate *risk level*, a *site survey*, and the definition of *site-specific design requirements* for the building envelope. The four-step process is described below, followed by examples of glazing systems from completed projects that were designed to resist extreme windstorms.

8.2   Buildings of special importance or with special functions

Building codes classify buildings according to occupancy and use. Buildings classified as 'important' or 'essential' facilities are assigned an 'importance factor' that makes them stronger by increasing the loads used in their design. The amount of this increase is, like the minimum design loads, also a building code minimum. As such, it may not be large enough to satisfy specific objectives for the preservation of building function. For example, the importance factor for important and essential facilities cited in the US standard for minimum design loads for buildings (ASCE, 2005) is 1.15 and is used to modify calculated design loads. When applied in accordance with provisions of the standard that address wind loads, this adjustment represents only a 7% increase in the design wind speed. This increase may appear inadequate when a facility is located in a hurricane or tornado prone area and the preservation of building function under extreme wind environments is paramount.

Public facilities such as hospitals, storm shelters, and schools are easily identified as being important or essential. To increasing degrees, however, commercial facilities such as financial centers, operations centers, and buildings housing important or valuable functions or inventories are being singled out as also requiring special attention. Preservation of building function, during and/or after a windstorm, has become as important to the design process as has the protection of its occupants.

8.3   Wind analysis

The risk of extreme winds at a building site is related to the wind climate in the region. Wind climates fall into three general categories: (1) wind regions outside areas subject to the occurrence of tropical cyclones, (2) wind regions subject to the occurrence of tropical cyclones, and (3) wind regions outside tropical cyclone prone areas where the occurrence of tornadoes may influence the risk analysis of extreme winds in a site-specific study. Wind risk is commonly expressed in terms of mean recurrence intervals (MRI) in which specific wind speeds will be exceeded, on average, over long periods of time. The MRI is the inverse of the probability of occurrence in the associated time interval.

8.3.1   Nontropical cyclone regions

For wind regions outside areas subject to the occurrence of tropical cyclones the wind risk may be defined using local wind speed records. The US national standard for defining minimum wind loads on buildings, ASCE

Table 8.1 Sample wind speed record in standard format. Annual extreme wind speed from local weather station (sample)

Year	Wind speed	Year	Wind speed	Year	Wind speed	Year	Wind speed
1	67	9	58	17	70	25	57
2	57	10	57	18	88	26	44
3	78	11	59	19	65	27	78
4	57	12	61	20	66	28	65
5	99	13	66	21	58	29	56
6	60	14	78	22	65	30	75
7	65	15	74	23	49	31	54
8	56	16	55	24	60	32	56

Wind speeds are annual extreme, 3 s gusts in miles per hour (mph) in open terrain at 10 m height. Sample mean $\bar{X} = 63.94$ and standard deviation $s = 11.34$.

7-05 (ASCE, 2005), allows use of basic wind speeds determined from regional climatic data if approved statistical-analysis procedures have been employed and if the length of record, sampling error, averaging time, anemometer height, data quality, and terrain exposure of the anemometer have been taken into account. Simiu and Scanlan (1996) present a procedure for probabilistic modeling of largest yearly wind speeds that meets these requirements. This procedure may be used to define the nontropical cyclone wind risk at a building site. Wind speed records are obtained from the nearest local weather station and converted to a standard format, commonly 3-second (3 s) gust wind speeds at a 10 meter (m) height in open (airport) terrain. A sample wind speed record in this format is presented in Table 8.1. This sample contains 32 years of data, expressed as the annual extreme wind speed for each year. The mean $\bar{X}$ and standard deviation s for the wind speed record are calculated and are shown in the table. The estimated value $\breve{v}_{\bar{N}}$ of the $\bar{N}$-year wind speed $v_{\bar{N}}$ is based on a Fisher–Tippett extreme value probability distribution and is expressed by

$$\breve{v}_{\bar{N}} = \bar{X} + 0.78(\ln \bar{N} - 0.577)s \qquad [8.1]$$

where $\bar{X}$ and s are, respectively, the sample mean and the sample standard deviation of the largest yearly wind speeds (annual extreme) from the sample record. The resulting wind risk for the sample data in Table 8.1 is presented in graphical form in Fig. 8.1.

8.3.2 Tropical cyclone regions

The procedure outlined above cannot be employed in wind regions subject to the occurrence of tropical cyclones because wind records at local weather stations will not contain enough tropical storm related wind speed data to make the process viable. In these regions ASCE 7-05 allows use of design

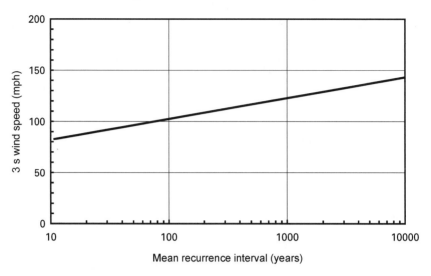

8.1 Wind risk model for nontropical cyclone region.

wind speeds derived from simulation techniques when approved simulation and extreme value statistical analysis procedures are employed. A wind risk model for hurricane winds that employs hurricane simulations is described in Vickery *et al.* (2000a, 2000b) and is combined with a filling (storm weakening) model described in Vickery (2005). Predictions of wind speed versus return period are provided for both surface level and upper level winds and are based upon a simulation of 100 000 years or more of tropical storms occurring in the Atlantic basin. The simulation methodology replicates tropical storms that have the same statistical properties as tropical storms from historical records that affected the site. In the simulation procedure wind speeds at a building site are recorded if simulated storms pass within 155 mi (250 km) and if the peak gust wind speed at the site exceeds 20 mph (8.9 m/s). Wind speeds and directions are computed every 15 minutes and maximum wind speeds produced by each simulated storm in each of 16 directional sectors are saved for use in an extreme value analysis. The simulation procedure also models changes in the hurricane boundary layer as a function of wind direction and the fetch distance from the coast. Predicted wind speeds at a site are derived by rank ordering the wind speeds resulting from the simulation of the 100 000 + years of storms. An interpolation technique is then used to obtain wind speed exceedance probabilities. The wind risks that are defined using this procedure are shown as conditional hurricane wind speed exceedance probabilities in Fig. 8.2 for an example building site in Florida.

General discussions of hurricane simulation techniques and other methods for assessing the risk of hurricane winds are contained in a paper

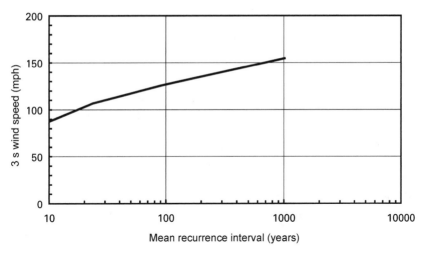

8.2 Wind risk model for tropical cyclone region.

by Vickery *et al.* (2007). In addition to the wind analysis method by Vickery *et al.* (2000a, 2000b, 2005) discussed above, procedures for assessing hurricane wind risks are offered by Emanuel *et al.* (2006), Georgiou *et al.* (1983), and Lee and Rosowsky (2007).

8.3.3 Tornado prone regions

Tornadoes factor into the wind analysis only in regions subjected to high rates of tornado occurrence and only for relatively large MRI. These regions are outside of hurricane prone regions; hence, the process for defining wind risk builds upon the procedure outlined in Section 8.3.1 for nontropical cyclone winds. Straight line winds govern at relatively low MRI and tornado winds govern at relatively large MRI. The level of risk at which tornado winds become a factor is related to the regional wind climate.

Estimating the probabilities of occurrence of tornado winds may follow a general procedure outlined by Simiu and Scanlan (1996). The probability that a tornado will strike a particular location in one year is $P(S)$. The probability that the maximum wind speed in a tornado will be higher than a value V_0 is $P(V_0)$. Multiplication of these two probabilities yields the probability that a tornado with maximum wind speeds higher than some specified value V_0 will strike a location in any one year $P(S,V_0)$. Values for $P(S)$ may be calculated using

$$P(S) = \bar{n}(\bar{a}/A_0) \qquad [8.2]$$

where $\bar{n}$ = average number of tornado occurrences per year

$\bar{a}$ = average individual tornado damage area

A_0 = area of tornado occurrences (e.g. a one-degree square)

An example by Simiu and Scanlan that outlines the procedure uses $\bar{n}$ = 2.06 (estimated from 13 years of tornado frequency data), $\bar{a}$ = 2.82 mi^2 (estimated for tornado occurrences in the State of Iowa), and A_0 = 4780 cos ϕ (in mi^2), where ϕ = 42.5° is the latitude at the center of the selected one-degree square. The calculation yields a tornado strike probability of 165 × 10^{-5} per year. An estimate of the probabilities $P(V_0)$ is based upon observations of 1612 tornadoes during 1971 and 1972 by Markee *et al.* (1974) and the rating of these tornadoes according to the tornado intensity scale by Fujita (1970). In this analysis the percent probability of tornadoes with wind speeds exceeding V_0 is

100 mph	50
150 mph	6.0
200 mph	2.2
250 mph	0.21
300 mph	0.09

Multiplication of these percentage values by the strike probability yields the annual probability of exceedance for each of these tornado wind speeds. The inverse of each wind speed probability is the MRI for that wind speed. Figure 8.3 illustrates the wind risk for a building site that contains both straight line winds derived from the procedure outlined in Section 8.3.1 and tornado wind speeds from the above example by Simiu and Scanlan. In this illustrative example tornado winds become a factor in the wind risk analysis at an MRI of about 3000 years.

The procedure by Simiu and Scanlan outlined above contains generalizations and subjective judgments, particularly with regard to the intensity of tornado winds and the areas that they affect. The procedure has been refined by focusing on the specific records of individual tornadoes within a selected one-degree square. Tornado records for a selected one-degree square in the United States may be obtained from the US National Climatic Data Center in Ashville, NC. The parameter $\bar{n}$ is refined, if necessary, by eliminating trends in occurrence data that may have been caused by reporting anomalies such as population growth. The parameter $\bar{a}$ is refined by classifying individual tornadoes by path length and width and using a methodology advanced by Abbey and Fujita (1979) to define subareas within the damage paths that were affected by specific intensities of tornado winds. $P(V_0)$ is refined to account for data anomalies, errors, and judgments in tornado intensity records. Twisdale (1978) and Twisdale *et al.* (1981)

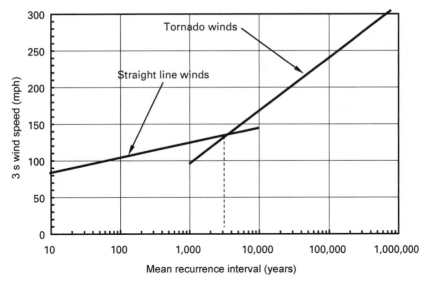

8.3 Wind risk model for tornado prone region.

address tornado intensity parameters, including temporal variations in tornado reporting frequency, errors in rating tornado intensities, under-rating of tornado intensity due to inhomogeneous distribution of buildings along the tornado path, and variation of tornado intensity along the tornado path. In the refined analysis $P(V_0)$ is expressed as probabilities of occurrence of specific wind speed ranges associated with comparable areas from the damage area ($\bar{a}$) calculations. Multiplications of $P(S)$ and $P(V_0)$ calculated in this manner yield the probabilities of occurrence of the specific wind speed ranges in the selected one-degree square. The inverse of these probabilities is the MRI.

8.4   Risk level

Perhaps the most difficult part of the process is the selection of an appropriate risk level. This decision is for the facility owner to make. It must balance the advantages of a more wind resistant structure or facility against the increased costs of providing added protection while meeting the objectives of the process. Oftentimes, costs are estimated for several risk levels to provide a cost/benefit analysis to aid in risk level selection. Fortunately, the process has been employed often enough for there to have been established a certain precedent. An evolving trend is to establish a risk level based on the planned operational status of the facility during and after an extreme wind event. Table 8.2 lists MRI for several important facilities for which risk analyses have been conducted and risk levels have been

Table 8.2 Risk levels. Selected risk level for important facilitites

Facility description	Special condition	Risk level (MRI)
Regional medical center	Surgical patients cannot be moved	500 years
Children's hospital	Infant intensive care unit	500 years
Local medical facility	Plan to remain open during storm	500 years
Local medical facility	Plan to close in advance of storm	300 years
International legal office	Function during and after storm	300 years

established. These examples use a 500 year MRI if the facility expects to remain functional during the storm and a 300 year MRI if the facility will close in advance of the storm, but is expected to be functional in the storm aftermath.

8.5 Site survey

An important aspect of the process for designing glazing systems to resist windstorms on special buildings is the character of potential windborne debris that may affect the facility. A facility located in open terrain or adjacent to bodies of water will experience less severe impacts from windborne debris than will a facility sited in an urban environment and/or surrounded by on-site infrastructure that has not been constructed to modern wind resistant standards. The design wind speed which, as noted above, is related to the risk of extreme winds, also plays into the character of the windborne debris that may impact the building façade. Hence, defining the specific characteristics of potential windborne debris becomes a site-specific undertaking that considers design wind speed (through the definition of risk level) and the local built environment.

Current building codes and standards and evolving precedent offer guidance to the windborne debris definition process. Minimum standards for windborne debris impacts are outlined in ASCE (2005), ICC (2006), and ASTM (2006). These standards prescribe a 'two-by-four' timber (a 2×4) weighing 9 lb (4 kg) and impacting at 50 ft/s (15 m/s) as the minimum design requirement for 'ordinary structures' in code-defined 'windborne debris regions'. 'Enhanced protection' for important structures requires resistance to the same 2 × 4 impacting at 80 ft/s (24 m/s) as a minimum standard. The upper bound of impacting debris is broadly defined by debris impacts specified for public shelters in FEMA (2000) and in DCA (2004). These standards prescribe a 15 lb (7 kg) 2 × 4 impacting at 73 ft/s (22 m/s). Debris impact requirements resulting from site surveys generally fall between these upper and lower bounds. Test methods for qualifying glazing systems under hurricane conditions also specify application of a regimen of 9000 cycles of wind pressure based on the design wind load following missile impact

Table 8.3 Windborne debris. Example windborne debris impact requirements

Wind speed	Site condition	Missile impact requirement
155 mph	Old buildings on site	15 lb 2 x 4 impacting at 65 ft/s
166 mph	Built-up urban area	15 lb 2 x 4 impacting at 65 ft/s
160 mph	Open/suburban site	9 lb 2 x 4 impacting at 80 ft/s
157 mph	Limited large missile potential	9 lb 2 x 4 impacting at 80 ft/s
152 mph	Isolated site near water	9 lb 2 x 4 impacting at 80 ft/s

(ASTM, 2006; ICC, 2006; FBC, 2004). Additional background on the character of windborne debris and its relationship to hurricane winds may be found in chronicles of the development of windborne debris impact standards by Minor (1994) and Hattis (2006) and in Chapter 7.

Precedent established in evolving practice also offers guidance for the selection of an appropriate debris impact requirement. The three major considerations are potential debris sources at the site, the risk level, and the associated design wind speed. Relatively open sites with lower risk levels and lower design wind speeds tend toward the lower bound of specified debris impacts. Large developments in which the building environment and infrastructure are controlled effectively through zone restrictions and ordinances may also justify lower bound debris impacts. Buildings sites in built-up areas where adjacent structures pre-date modern wind load standards suggest use of the more severe debris impacts, especially when design wind speeds are relatively large. Table 8.3 lists several windborne debris impact requirements from completed projects along with considerations employed in their selection.

8.6   Site-specific design requirements

Site specific design requirements for design of the glazing systems to resist extreme windstorms are expressed in terms of a selected level of risk, a design wind speed, and a specified requirement for debris impact. The selected *risk level* from Section 8.4 (expressed in terms of MRI or probability of occurrence) dictates the design wind speed through the *wind analysis* (one of the risk models in Section 8.3). Debris impact requirements are defined according to the *site survey* considerations discussed in Section 8.5.

8.7   Design examples

Glazing system designs may be based upon previously employed and tested systems such as those described below or from new glazing systems designed and tested according to the provisions of applicable building codes and standards. In the latter case, the requirements for design are advertised and tenders are invited from specialty glazing system suppliers. The process described in this chapter has been employed in the design of glazing systems for a wide range of public and commercial facilities. In addition to hospitals and public shelters, the process has been applied to the design of financial data centers, cruise line terminals, airline operations centers, telecommunications centers, international law offices, large mixed-use developments, and truck dispatching centers. In many of these design situations the openness of the glazing system for the admission of daylight and the preservation of sightlines have not been compromised in order to achieve added protection.

8.7.1 Hospital

An addition to a regional medical facility in Georgia was to contain a surgical tower in which intensive care patients could not be relocated in advance of a hurricane. The wind risk was determined for the building site using the hurricane simulation procedure discussed in Section 8.3.2. The acceptable risk level was defined as a MRI of 500 years. A 3 s gust wind speed of 155 mph (69.3 m/s) was associated with this risk level through the wind risk analysis. This wind speed was then specified for the testing of glazing systems for impact resistance. A site survey revealed a predisposition for the generation of significant amounts of windborne debris from on-site buildings that had been built prior to the appearance of building codes that contain modern wind load provisions (c.1975). Hence, the specified design missile was a relatively severe 15 lb 2 × 4 timber traveling at 65 ft/s (20 m/s) to be tested in conformance with Dade County (Florida) protocols PA 201-94, PA 202-94, and PA 203-94 (SFBC, 1994). These test protocols are similar to those found in ASTM E1996-06. The glazing system that was selected and tested to meet this impact requirement was an $\frac{11}{16}$ in (17 mm) Sentryglas heat-strengthened glass product. The glazing detail is illustrated in Fig. 8.4. The expected performance of this product upon impact from the design missile assumes glass breakage and retention of broken glass particles by the polyvinyl butyral/polyethylene terephthalate (PVB/PET) laminate that is bonded to the inner glass surface. To prevent fallout of the broken glass/laminate system after impact and subsequent pressure cycles, the perimeter of the glass panel is anchored to the window frame by a silicone sealant with substantial glass contact width ($\sim \frac{3}{4}$ in; 18 mm) that acts as an anchor bead.

8.7.2 International legal office

A new building for a legal office in the Caribbean was planned to service an international clientele on an around-the-clock basis. A site-specific wind analysis was conducted and a 300 year MRI was selected as an appropriate risk level to assure a suitable level of availability during and after a

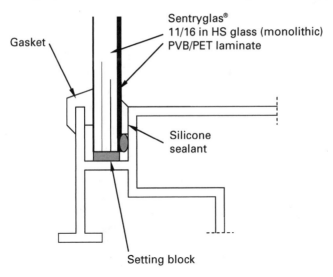

8.4 Glazing detail for Georgia hospital.

hurricane. The associated wind speed was 156 mph (69.7 m/s). Adjacent urban buildings portended significant levels of windborne debris; hence, a 15 lb 2 × 4 timber traveling at 65 ft/s (20 m/s) was selected as the design missile. The glazing system that was selected and tested to meet this impact requirement was a laminated glass/polycarbonate system consisting of a polycarbonate core, adjacent polyurethane layers (to accommodate differences in thermal expansion), and heat-strengthened glass exterior plies. The glazing detail is illustrated in Fig. 8.5 and a photograph of the building which was designed by Chalmers Gibbs Architects, Grand Cayman Islands, is shown in Fig. 8.6. The expected performance of this product upon impact from the design missile assumes breakage of one or both of the glass plies, but that the polycarbonate core will not fracture. Hence, the glazing system is held in place by the stiffness of the core ply. This system may be 'dry glazed'; i.e. there is no requirement for the silicone anchor bead illustrated in Fig. 8.4 to hold the impacted system in place. The product manufacturer recommends, however, perimeter seals to protect the glass/polycarbonate edges from moisture intrusion.

8.7.3 Level E glazing system

ASTM E1996 specifies a missile impact requirement for 'enhanced protection' in hurricane wind zones where the 3 s design wind speed exceeds 130 mph (58.1 m/s). This missile is a 9 lb (4 kg) 2 × 4 timber impacting at 80 fps (24 m/s) and is designated as a Level E missile. The Office of Code Compliance in Dade County (Miami), Florida, conducts a

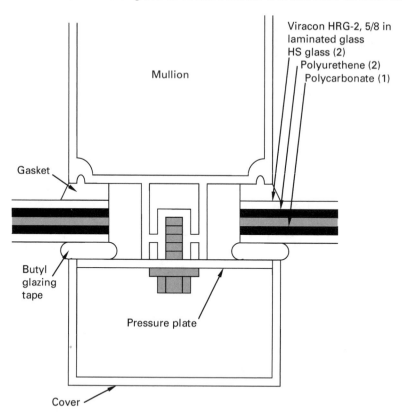

8.5 Glazing detail with polycarbonate core laminated glass.

8.6 Ugland House, George Town, Grand Cayman Islands (photograph courtesy of Jonathan Ashton, Chalmers Gibbs Architects).

product approval program and maintains a register of products that have been tested in accordance with hurricane related standards in the *Florida Building Code* (FBC, 2004) as well as in other building codes and standards. The insulating glass (IG) glazing system illustrated in Fig. 8.7 has been accepted as a tested system that meets the ASTM E1996 Level E missile impact requirement. Two laminated glass (LAG) lites constitute the IG unit which is anchored to the framing system by a silicone anchor bead. Each LAG lite has a special interlayer. The outer lite is comprised of two $\frac{1}{4}$ in (6 mm) heat-strengthened glass plies separated by a 0.075 in (2 mm) Vanceva™ PVB–PET–PVB interlayer by Viracon. The inner lite is comprised of two $\frac{1}{4}$ in (6 mm) heat-strengthened glass plies separated by a 0.10 in (3 mm) enhanced PVB interlayer by Solutia. The glazing system is manufactured by Hurricane Manufacturing Company. Upon impact by the specified missile both lites break, but the missile does not penetrate the glazing. The broken lites are held in the opening by the silicone anchor bead during subsequent pressure cycling. This Level E window system was designed and fabricated by Hurricane Manufacturing Corporation, Inc. and installed in a building owned by Sarasota County, Florida that houses the Florida Department of Health (Fig. 8.8).

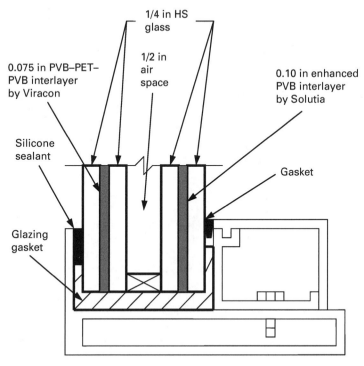

8.7 Glazing detail for Level E missile impact.

8.8 Sarasota County, Florida Department of Health (photograph courtesy of Jeff Robinson, Hurricane Manufacturing Corporation, Inc.).

8.8   Conclusions

The visibility of hurricanes and other extreme natural events and increasing concerns with the impacts of these events on public infrastructure, emergency operations, and industrial activity have fostered new concerns for the built environment. No longer is it sufficient to design a building or facility for life safety only. Many essential and important buildings and facilities must operate through and following an extreme event with a minimum of disruption. Where extreme wind events such as hurricanes and tornadoes are involved, the conventional design process has been expanded. The expanded procedure includes a *wind analysis*, the selection of an appropriate *risk level*, a *site survey*, and the definition of *site-specific design requirements* for the building envelope. This process is becoming established as a new standard of practice in which guidelines and precedent are evolving that will assist designers of these facilities.

8.9   References

Abbey, R. F. and Fujita, T. T., (1979) The dapple method for computing tornado hazard probabilities: refinements and theoretical considerations, in Proceedings of the Eleventh Conference on *Severe Local Storms*, Kansas City, October 1979, American Meteorological Society, Boston, MA.
ASCE (2005) ASCE/SEI 7-05, *Minimum Design Loads for Buildings and Other Structures*, American Society of Civil Engineers, Reston, VA.
ASTM (2006) ASTM E1996-06, *Performance of Exterior Windows, Glazed Curtain Walls, Doors and Storm Shutters Impacted by Windborne Debris in Hurricanes*, American Society for Testing and Materials, West Conshohocken, PA.
DCA (2004) *State of Florida 2004 Statewide Emergency Shelter Plan*, Florida Department of Community Affairs, Tallahassee, FL.

Emanuel, K. A, Ravela, S., Vivant, E. and Risi, C. (2006) A statistical–deterministic approach to hurricane risk assessment, *Bulletin of American Meteorological Society*, 19, 299–314.

FBC (2004) *Florida Building Code*, Florida Building Commission, Department of Community Affairs, Tallahassee, FL.

FEMA (2000) FEMA 361, *Design and Construction Guidance for Community Shelters*, Federal Emergency Management Agency, Washington, DC, July 2000.

Fujita, T. T. (1970) *Proposed Characterization of Tornadoes and Hurricanes by Area and Intensity*, Research Paper No. 89, Satellite and Mesometeorology Research Project, University of Chicago.

Georgiou, P. N., Davenport, A. G. and Vickery, B. J. (1983) Design wind speeds in regions dominated by tropical cyclones, in Proceedings of the Sixth International Conference on *Wind Engineering* Gold Coast, Australia, 21–25 March 1983 and Auckland, New Zealand, 6–7 April 1983 Australasian Wind Engineering Society.

Hattis, D. B. (2006) Standards governing glazing design in hurricane regions, *Journal of Architectural Engineering*, 12(3), September 1, 2006, 108–115.

ICC (2006) *International Building Code*, International Code Council, Falls Church, VA.

Lee, K. H. and Rosowsky, D. V. (2007) Synthetic hurricane wind speed records: development of a database for hazard analyses and risk studies, *Natural Hazards Review*, 8, 23–34.

Markee, E. H., Beckerley, J. G. and Sanders, K. E. (1974) *Technical Basis for Interim Regional Tornado Criteria*, WASH 1300 (UC-11), US Atomic Energy Commission, Office of Regulation, Washington, DC.

Minor (1994) Windborne debris and the building envelope, *Journal of Wind Engineering and Industrial Aerodynamics*, 53, 207–227.

SFBC (1994) Metropolitan Dade County Edition of the *South Florida Building Code*, Metropolitan Dade County, Miami, FL.

Simiu, E. and Scanlan, R. H. (1996) *Wind Effects on Structures*, 3rd edn, John Wiley & Sons, Inc., New York.

Twisdale, L. A. (1978) Tornado characterization and wind speed risk, *Journal of Structural Division, ASCE*, October 1978, 1611–1630.

Twisdale *et al.* (1981) *Tornado Missile Simulation and Design Methodology*, EPRI NP-2005, Electrical Power Research Institute, Palo Alto, CA, August.

Vickery, P. J. (2005) Simple empirical models for estimating the increase in the central pressure of tropical cyclones after landfall along the coastline of the United States, *Journal of Applied Meteorology*, 44(12), 1807–1826.

Vickery, P. J., Masters, F. J., Powell, M. D. and Wadhera, D. (2007) Hurricane hazard modeling: the past, present and future, in Proceedings of the Twelfth International Conference on *Wind Engineering*, Cairns, Australia, July 1–6, 2007, Australasian Wind Engineering Society.

Vickery, P. J., Skerlj, P. F., Steckley, A. C. and Twisdale, L. A. (2000a) Hurricane wind field model for use in hurricane simulations, *Journal of Structural Engineering*, 126 (10), 1203–1221.

Vickery, P. J., Skerlj, P. F. and Twisdale, L. A. (2000b) Simulation of hurricane risk in the U.S. using empirical track model, *Journal of Structural Engineering*, 126 (10), 1222–1237.

9

Test methods for performance of glazing systems and exterior walls during earthquakes and extreme climatic events

S. A. WARNER Architectural Testing Inc., USA

Abstract: This chapter explains the importance of building and testing exterior wall mock-ups prior to the start of a construction project. With a rise in awareness of the need to improve the overall building performance, standards writing organizations have established specifications and test procedures designed to address some of the most vulnerable construction elements, namely exterior walls and roofs. This chapter outlines the purpose of preconstruction mock-up testing, introduces a standard testing sequence, and then provides an overview of each of the test protocols developed for use by the industry.

Key words: mock-up testing, air/water/structural performance testing, exterior wall test specimen, static pressure, dynamic pressure, seismic-induced building movement, wind-induced building movement, interstory building movement, thermal expansion and contraction, serviceability testing.

9.1 Introduction

Testing of exterior wall mock-ups has become an important and standard practice for architects, owners, and building contractors. The reasons for the increased frequency of mock-up testing include specification/code compliance, owner/contractor due diligence, as well as risk management. Also, consideration is often given to the concern for building performance during catastrophic events such as those occurring at earthquake and hurricane prone regions. Earthquakes caused major damage in Northridge, California (1994), and Kobe, Japan (1995). Recently, the 2004 and 2008 hurricane seasons in the United States were especially severe. The four major hurricanes that struck the southeastern United States in 2004 were Category

2, 3, and 4 storms. Hurricane Charlie was the first Category 4 storm since Andrew in 1992. Hurricane Frances (Category 2), Hurricane Jeanne (Category 3), and Hurricane Ivan (Category 3) all caused significant damage to buildings across the southeast.

As reported by the Federal Emergency Management Agency (FEMA) in FEMA 490, *Summary Report on Building Performance*, the year 2004 set a record for the number of presidential disaster declarations in the United States (27 of the 68 disaster declarations were due to hurricanes). These severe storms often cause serious damage to buildings, especially the exterior walls and roofs, which are often the most vulnerable construction elements.

The FEMA report emphasizes the need for improvement in overall building performance. Industry and standards writing organizations, such as the American Architectural Manufacturers Association (AAMA) and the American Society for Testing and Materials (ASTM), have responded to the need for enhanced testing methods and specifications for the building envelope. AAMA publishes over 150 documents related to testing of fenestration products and components while ASTM publications are divided into 80 separate volumes with more than 12 000 individual standards. Over 1300 of the ASTM standards are referenced in the US building codes, of which many relate to the exterior building envelope. The model building code is a *minimum* design standard related primarily to protecting life safety. Most project specific test specifications for glazed walls go well beyond code minimums ('code plus') and emphasize serviceability and durability of glazing systems and building envelope wall systems.

The following is a summary of the commonly used AAMA and ASTM methods related to glazing systems and preconstruction exterior wall mock-ups:

- AAMA 501-05, *Methods of Tests for Exterior Walls*, is a comprehensive compilation of AAMA-recommended tests. Originally released by AAMA in 1968, it was updated in 1983, 1994, and again in 2005. AAMA 501-05 is not specifically referenced by the IBC; however, the document does provide a guide specification for mock-up testing of exterior walls.
- AAMA 501.1, *Standard Test Method for Exterior Windows, Curtain Walls and Doors for Water Penetration Using Dynamic Pressure,* uses a propeller aircraft engine or similar wind generator to create a wind-driven rain test condition.
- AAMA 501.2 and 501.3 are field testing documents for the in situ evaluation of air leakage and water penetration resistance. These documents are not reviewed in this chapter.
- AAMA 501.4-2000, *Recommended Static Test Method for Evaluating Curtain Wall and Storefront Systems Subjected to Seismic and Wind*

Induced Interstory Drifts, is a static racking test method focusing on the air and water serviceability of a curtain wall or a storefront system after exposure to simulated static seismic movement.

- AAMA 501.6-2001, *Recommended Dynamic Test Method for Determining the Seismic Drift Causing Glass Fallout from a Wall System Panel,* in contrast to AAMA 501.4 above, is a dynamic racking test method focusing on the seismic safety of architectural glass components and cladding within a curtain wall or a storefront system. The AAMA 501.6 document addresses concerns for falling shards of glass resulting from seismic-induced building movements. Until recently, the performance of buildings enduring seismic effects has been virtually ignored in building codes. Now, thanks to the efforts of concerned engineers, architects, and designers in cooperation with AAMA, both the AAMA 501.4 and 501.6 seismic test methods are referenced in the building codes and in ASCE 7.

- AAMA 501.5, *Test Method for Thermal Cycling of Exterior Walls,* provides a standardized procedure to expose mock-ups to temperature extremes followed by air and water leakage serviceability testing. Thermal cycling is an important indicator of long-term performance and is often a required preconstruction mock-up test specified by architects.

In addition to the AAMA testing standards referenced above, ASTM has published several new documents for the evaluation of glazed wall systems and exterior wall mock-ups. ASTM E 1886, E 1996, and E 2099 have all been published in recent years. ASTM E 1886 and E 1996 relate to hurricane-resistant glazing. ASTM E 2099 is a standard practice for evaluating mock-ups and provides a default testing sequence for testing. Historically, ASTM E 283, E 331, and E 330 form the foundation for most exterior wall mock-up testing programs. Commonly referred to as the air, water, and structural test series, ASTM originally published these documents in 1964 and has continued to update and maintain them. The air/water test series is often repeated after other performance testing protocols such as thermal and building movement in order to quantify the serviceability impact to the wall.

9.2   The purpose of testing exterior wall mock-ups

The curtain wall industry has successfully used preconstruction mock-up testing for more than sixty years to provide evidence that walls comply with specified standards. Today preconstruction mock-up testing includes all types of wall and roof assemblies. Testing of exterior mock-ups has evolved from this basic preconstruction screening test to testing for structural adequacy, along with the weather tightness, durability, and life safety attributes of the wall. Mock–up testing is now the rule rather than the exception. This change is due in part to the complexity of wall designs but is also driven by design and construction related litigation as well as increased building code enforcement by many jurisdictions.

Modern performance standards for exterior walls are more robust and wall providers often perform tests as an integral part of the design and development process. Although it is impossible to provide a 100 % guarantee that a wall will perform in all possible environmental conditions on a given building, a well-planned and executed preconstruction mock-up testing program often reveals design weaknesses, workmanship problems, and fabrication defects at a point in the process at which they can be addressed with minimal impact on the project's schedule and cost.

9.3   The exterior wall test specimen

The exterior wall test specimen must be as close as possible to a faithful representation of the intended design, including supports, glazing, anchors, sealants, and flashing details. Similarly, the installation practices used should be the same as those to be used on the project. If possible, it is best to have the personnel who supervise or erect the mock-up to also install (or at a minimum supervise the installation of) the actual wall on the building.

The size and configuration of the mock-up are critical decisions that require careful consideration by the design professional. Corner elements, transitions areas, and intersecting areas of wall elements should be included whenever possible in a comprehensive mock-up assembly. It is essential that the preconstruction mock-up be of sufficient height (a minimum of one typical floor height plus additional height so that all typical horizontal conditions are represented) and a minimum width of two repetitive bays plus one additional bay so that all typical vertical members are represented.

9.4   The testing sequence

The most common basic mock-up testing sequence is as follows:

1. Preload per ASTM E 330 at 50 % of the specified positive design wind pressure.
2. Air leakage per ASTM E 283.
3. Static water resistance per ASTM E 331.
4. Dynamic water resistance per AAMA 501.1.
5. Structural performance at design loads per ASTM E 330.
6. Structural performance at 150 % of design loads per ASTM E 330.

A more comprehensive testing sequence is often dictated by project-specific specifications and includes optional tests as shown below:

1. Preload per ASTM E 330 at 50 % of the specified positive design wind pressure.
2. Air leakage per ASTM E 283.
3. Static water resistance per ASTM E 331.
4. Dynamic water resistance per AAMA 501.1.
5. Structural performance at positive and negative design wind load per ASTM E 330.
6. Repeat air leakage per ASTM E 283.
7. Repeat static water penetration resistance per ASTM E 331.
8. Repeat dynamic water resistance per AAMA 501.1.
9. Static seismic, wind, and floor slab movement per AAMA 501.4.
10. Repeat air leakage per ASTM E 283.
11. Repeat static water penetration resistance per ASTM E 331.
12. Repeat dynamic water resistance per AAMA 501.1.
13. Thermal cycling test per AAMA 501.5.
14. Repeat air leakage per ASTM E 283.
15. Repeat static water penetration resistance per ASTM E 331.
16. Repeat dynamic water resistance per AAMA 501.1.
17. Structural performance at 150 % of design wind loads per ASTM E 330.
18. Static seismic movement at 150 % of the specified lateral movement per AAMA 501.4.
19. Dynamic seismic tests per AAMA 501.6 (these tests require a separate mock-up; refer to Section 9.11 for a detailed explanation).

All of the above tests are described more fully in Sections 9.5 to 9.12 below.

9.5   Air leakage (ASTM E 283)

The ability of a wall system to control air infiltration and exfiltration is important to the proper functioning of the building's mechanical systems. Air leakage is also important with regard to the energy consumption, durability, and condensation resistance performance of a building. Window and door products are tested and certified by the manufacturer for air leakage resistance in accordance with the National Fenestration Rating Council's NFRC 400 or AAMA 101/I.S.2/ A440. Air barrier materials are tested and prequalified for air leakage per ASTM E 2178.

The rate of air leakage for the exterior wall mock-up is determined under specified pressure differential conditions across the test specimen. The most commonly used pressure differentials are 1.57 psf (75 Pa) and 6.24 psf (300 Pa). These pressure differentials are roughly equivalent to 25 mph (40 km/h) and 50 mph (80 km/h) wind speeds. The total measured air flow at the specified test conditions is a combination of the mock-up specimen air

9.1 A 4 mil plastic film is applied over the exterior face of the preconstruction mock-up as required to determine accurately extraneous chamber leakage (tare reading) during the air leakage test.

leakage and the test chamber air leakage. Since it is usually impractical to eliminate all of the test chamber leakage it must be measured by sealing the specimen and repeating the test. The extraneous air leakage (chamber tare) is accurately measured by placing a film over the exterior face of the mock-up and subjecting the specimen to the specified pressure differential (see Fig. 9.1).

9.6   Tests for water penetration using static pressure (ASTM E 331)

The resistance to static water infiltration is measured by applying water to the exterior face of the mock-up at the rate of 5 gallons per square foot per hour (3.4 L/m^2 min) while subjecting the mock-up to the specified pressure differential (see Fig. 9.2). The most commonly specified pressure differentials for the static water penetration resistance test range from 6.24 psf (300 Pa) to 15.0 psf (720 Pa). These test pressures approximate wind driven rain pressures of 50–75 mph (80–120 km/h).

Obviously, selection of an appropriate wind-driven rain water test pressure requires an understanding of the forces imparted to the building envelope during these environmental conditions. Wind pressure and kinetic energy can force water through openings in the building envelope or overload a glazing system's water retention gutters.

As noted in AAMA/WDMA/CSA 101/I.S.2/A440,

9.2 Static water resistance testing on a full-scale preconstruction mock-up. Water is applied to the exterior face at the rate of 5.0 US gallons per square foot per hour.

... Three things are required to move water through a surface, a source of water, a path for the water to follow, and a force to drive the water through the opening. If any one of these items is absent, leakage cannot occur ... The forces, which can drive leakage, are generally considered to be kinetic forces, gravity, capillary action, surface tension, and pressure differentials. In some circumstances only one or two of these forces may be present, but in a windy rainstorm all of them will likely be acting to move the water through any available leakage path .... A pressure difference can drive water through any small leakage paths including those having a limited upward slope. The direction of the flow is from the side with higher pressure to the side with lower pressure. This Standard/ Specification requires that the minimum water penetration resistance test pressure be determined as a percentage of the positive design pressure (DP) because this condition renders the biggest pressure difference between internal pressure of the building, external wind pressure, and the conditions to drive water to the interior of the building.

Less obviously, strong winds have been shown to affect water droplet size and sheeting action on the outside of a building. Research into the complex physical behavior of wind-driven rain in hurricane conditions continues, supported by AAMA as well as other interested agencies and organizations.

While 3 second gusts referenced in ASCE 7 are the standard for structural design, they may not be suitable for establishing an appropriate wind-driven rain resistance test pressure. The selection of an appropriate wind-driven rain pressure is predicated on wind and rain events occurring at the same time. ASCE 7 methods do not assume rain events coupled with wind. ASTM E 331 tests are conducted at a constant static pressure maintained for a period of 15 minutes. Non-pressure-equalized systems and components do not 'fill' to their water head equilibrium height immediately, but rather 'build' water as the test proceeds. Most are at equilibrium 5 to 10 minutes into the 15-minute test duration. (By nature, pressure-equalized systems reach equilibrium quicker, but seldom reach the maximum water head height.).

For example, at the ASCE 7 basic wind speed 90 mph contour, the 5 minute (300 second) velocity is approximately 65 mph. This wind velocity creates less than a 12 psf inward-acting pressure for a building with a mean roof height of 150 feet in Exposure C, using Equation 6.15 from ASCE 7-05. The same exposure results in an inward-acting pressure of approximately 10 psf for a low-rise residential structure of roof height less than 60 ft. It is apparent from this simple example that the selection of appropriate project-specific wind-driven rain pressures is not an exact science. It is expected that test methods, standards, and specifications related to wind-driven rain performance of buildings will be enhanced and updated as additional research is made public.

9.7   Tests for water penetration using dynamic pressure (AAMA 501.1)

The resistance to water infiltration due to dynamic pressure is measured by applying water to the exterior face of the mock-up at the rate of 5 gallons per square foot per hour ($3.4 \ \text{L/m}^2$ min) while subjecting the mock-up to the specified dynamic wind pressure (see Figs 9.3 and 9.4). The selection of the appropriate dynamic wind pressure is similar to the discussion regarding static water test pressure. The design professional must consider the building exposure along with the risk tolerance of the owner when establishing the test pressure for a specific project. In the dynamic water test, a turbulent air flow is directed against the wall simultaneous with the application of water to the exterior face. Similar to the static test the most common pressure differentials

9.3 Airplane engine is used to create dynamic pressure for the wind-driven rain test.

9.4 Airplane engine is used to create dynamic pressure for the wind-driven rain test. The dynamic wind speeds range from 50 to 80 mph for most dynamic tests (Figure 1 of the typical AAMA 501.4 test configuration).

for this test also range from 6.24 psf (300 Pa) to 15.0 psf (720 Pa). The engine rpm's required to produce the equivalent wind velocities are pre-recorded for each of the prescribed test pressures during a calibration sequence. Water is introduced into the wind stream at the specified rate. The turbulence forces created during the dynamic water test may reveal sources of water penetration that the uniform static air pressure test (ASTM E 331) would not reveal. The dynamic water test method is considered by many experts to be more representative of a severe wind-driven rain event, which often creates unpredictable and suddenly shifting wind gusts and wind-blown water.

9.8   Static tests for seismic-induced building movement (AAMA 501.4)

This test uses static racking at a design-specific horizontal displacement or a default value of 0.01 multiplied by the story height. It is a static racking test focusing on the serviceability of curtain wall systems or storefronts during earthquakes. Thus, testing is conducted on a full-scale, multistory mock-up to determine the ability of the curtain wall or storefront to withstand a specified design displacement. This testing is followed by system service-ability tests for air and water infiltration control (see Fig. 9.5).

AAMA 501.4 provides the pass/fail criteria for three different types of facilities: (1) essential; (2) high-occupancy assembly; and (3) standard occupancy. The pass/fail criteria include provisions for functionality and

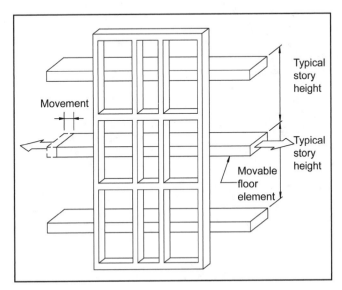

9.5 Typical AAMA 501.4 test configuration (reproduced by permission of American Architectural Manufacturers Association)

visible damage, glass breakage, and post displacement performance parameters.

9.9   Test for differential movement due to thermal expansion and contraction (AAMA 501.5)

This test is conducted to evaluate an exterior wall system's ability to maintain weather tightness (air leakage and water penetration) after exposure to a specified number of thermal cycles. The test is performed by covering the outdoor side of the test mock-up with an insulated enclosure

9.6 Three-story (32 feet) insulated enclosure is used for thermal cycling of a large preconstruction mock-up with outside 90° corner condition.

equipped with a means to raise or lower the exterior ambient temperature (see Fig. 9.6).

9.10 Tests for structural performance (ASTM E330)

Structural performance tests are commonly conducted at 50 %, 100 %, and 150 % of the positive and negative design wind pressures (see Fig. 9.7). The determination of appropriate design wind pressure is usually the responsibility of the building designer or engineer of record. Wind provisions in US Standards and Codes are governed by the American Society of Civil Engineers (ASCE). The 2006 IBC has adopted ASCE 7-05 for wind design.

It is typical that once a specimen has passed the tests at the specified design pressure, the test pressure is increased until failure occurs (see Fig. 9.8). Two procedures can be used for conducting structural tests; however, both procedures require deflection measurements of the glass supporting frame members. The first is used when a load–deflection curve is not required. Here, the test specimen is merely subjected to the specified test load. The second procedure is used when a load–deflection curve is required. Generating a load–deflection curve is generally reserved for research and development testing; therefore, the first procedure is the default for project specific mock-up testing.

If glazing breakage occurs before the specified maximum test pressure is achieved, it is necessary to examine the mock-up to determine the cause of failure. If the glazing breakage is determined to be the result of the

9.7 Dial indicators attached to mullions of a curtain wall system to monitor and record deflection at design wind pressure.

9.8 Glazing failure in performance during structural testing. Notice the 'wet blanket' effect of a fully tempered laminated glass unit.

interaction with the perimeter framing members (i.e. excessive deflections, failure of fasteners, etc.) the test is terminated and revised design considerations may be required. If the cause of the glazing breakage cannot be determined the glazing may be replaced and the test is repeated. Some specifications limit the number of glazing failure and repeat tests.

9.11   Dynamic seismic tests (AAMA 501.6)

AAMA 501.6-2001, *Recommended Dynamic Test Method for Determining the Seismic Drift Causing Glass Fallout from a Wall System Panel,* is a dynamic racking test method focusing on the seismic safety of architectural glass components within curtain and storefront wall systems. Essentially, the AAMA 501.6 test involves mounting individual, fully glazed wall panel specimens on a dynamic racking test apparatus (Fig. 9.9), which moves back

9.9 Schematic rendering of the AAMA 501.6 racking test facility.

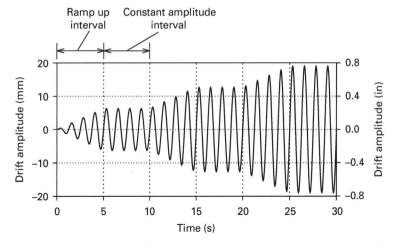

(a) First 30 seconds of crescendo test

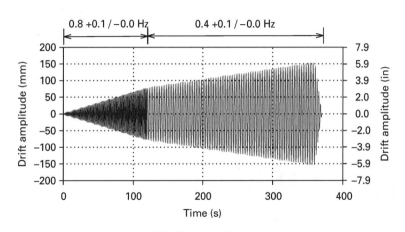

(b) Full crescendo test

9.10 Schematic of displacement time history for the dynamic crescendo test.

and forth horizontally in sinusoidal motions at gradually and progressively higher racking amplitudes, as in a musical crescendo (Fig. 9.10).

Dynamic racking frequencies are 0.8 Hz at lower racking amplitudes ($< \pm 75$ mm (3 in)) and 0.4 Hz at higher racking amplitudes ($> \pm 75$ mm). The racking amplitude at which glass fallout is first observed for a given specimen is designated as $\Delta_{fallout}$ for that test specimen. The lowest value of racking displacement causing glass fallout for the three specimens tested in accordance with AAMA 501.6 is the reported value of $\Delta_{fallout}$ for that particular wall system glazing configuration. This value of $\Delta_{fallout}$ is used in Equation 9.6.2.10.1-1 from ASCE 7-02 seismic design provisions:

$\Delta_{\text{fallout}} \geq 1.25\ ID_p$ (Equation 9.6.2.10.1-1 per ASCE 7-02)
or 13 mm (0.5 in), whichever is greater

where

Δ_{fallout} = relative seismic displacement (drift) causing glass fallout from the curtain wall, storefront wall or partition (Section 9.6.2.10.2)

D_p = relative seismic displacement that the component must be designed to accommodate (Equation 9.6.1.4-1), which shall be applied over the height of the glass component under consideration

I = occupancy importance factor (see Table 9.1.4 from ASCE 7–02)

AAMA 501.6 is intended to supplement the AAMA 501.4 standard and uses representative architectural glass panels as the test specimens (see Fig. 9.11). The goal of AAMA 501.6 is to determine the horizontal racking displacement required to reach the ultimate limit state of a glazing product (i.e. glass fallout). This dynamic racking test is performed until failure occurs, whereas AAMA 501.4 is performed according to a predetermined static racking displacement limit. Glass fallout is defined to have occurred in AAMA 501.6 when an individual glass fragment larger than 1.0 in^2 (650 mm^2) falls from the glazed opening in any direction. The horizontal racking displacement at which this glass fallout occurs, Δ_{fallout}, is the racking displacement for which the ultimate limit state is reached for that particular glazing panel.

Three representative glazing panel samples are racked according to a

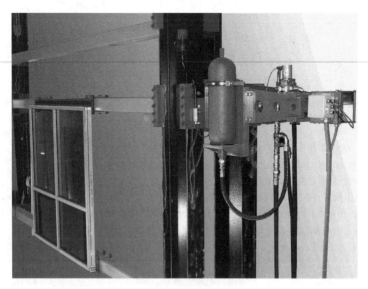

9.11 Dynamic seismic testing apparatus with one test sample installed.

9.12 Dynamic seismic testing apparatus with three test specimens installed.

crescendo test sequence and the amplitude of racking displacement required to cause fallout of each specimen is recorded (see Fig. 9.12). The smallest value of $\Delta_{fallout}$ observed in the three tests is the value of $\Delta_{fallout}$ to be used in the building code design equation for the seismic design of architectural glass.

9.12   Evaluation of mock-up test results

Given the ever growing complexity and variety of building envelope systems, the evaluation of their performance via preconstruction mock-up tests is critical to ensure performance during the service life of the building. If properly executed, standard laboratory tests can significantly reduce problems encountered in the field. It is not unusual for a wall mock-up to fail a test. When this happens, the remedy involves a forensic process to determine the origin and type of failure. Often this process requires isolation testing and partial disassembly of the mock up. Once the location of the failure is determined, it is important to characterize the failure as one of either design, material, or workmanship. It is important to include the entire construction team in the forensic process. Workmanship deficiencies must be flagged for close inspection during installation on the building. Material deficiencies must be evaluated by the responsible material suppliers and may require additional testing. Design deficiencies are often the most problematic and can severely impact the construction schedule. For this reason, it is very important that the exterior wall mock-up testing be scheduled well in advance of the schedule for delivery of materials to the project site. Often the architect will employ the services of an exterior wall consultant for this forensic process. Properly qualified exterior wall consultants can often locate the source of the failure and assist the construction team with appropriate corrective actions. Upon successful completion of the testing the wall contractors should provide 'as built' mock-up drawings showing any remedial or additions required to meet the specified performance. The as-built mock-up drawings are usually referenced by the testing laboratory and are often included as an attachment to the test report. The mock-up materials, design, and workmanship as tested provide the standard for field installation. At the conclusion of testing the contractor shall revise all shop drawings and resubmit to the architect for acceptance.

9.13   Meeting the building code

The building codes of most jurisdictions within the United States are based on a version of the *International Building Code* (IBC). The first IBC code was published in 2000. This consolidation in 2000 combined the *National Building Code* (NBC), the *Standard Building Code* (SBC), and the *Uniform Building Code* (UBC).

There are substantial advantages in combining the efforts of the existing code organizations to produce a single set of codes. Code enforcement officials, architects, engineers, designers, and contractors can now work with a consistent set of requirements throughout the United States. Manufacturers can put their efforts into research and development rather than designing to three different sets of building code provisions, and can focus on being more competitive in worldwide markets.

It is important to re-emphasize that compliance with the model building codes is a *minimum* design standard related primarily to protecting life safety. Most project-specific test specifications go well beyond code minimums, and emphasize long-term serviceability of glazing systems and building envelope wall systems. Specifically, the testing outlined below is code mandated for glazing and glazed exterior walls:

Code Section	Requirement	Reference
802.3.1 (IECC)	Air leakage	ASTM E 283
1609 (IBC)	Wind loads	ASTM E 330
2404 (IBC)	Seismic	ASCE 7
2404 (IBC)	Snow load	ASTM E 1300
2406 (IBC)	Safety glazing	ANSI Z97.1

9.14 The product/systems approval and certification process

For most commercial projects the product approval is accomplished via the contract submittal process. The product manufacturer is responsible for submitting test reports in accordance with the project specification. The architect or project consultant reviews and approves the submission.

Some glazing systems are submitted for approval to third party independent certification entities. The American Architectural Manufacturers Association has the oldest and most recognized such certification program in the United States.

Under the AAMA certification program a third party independent administrator verifies that the product meets the requisite tests according to applicable standards. Once all tests and conformance levels have been verified the administrator issues a Notice of Product Certification to the manufacturer, who then may purchase AAMA certification labels for application to product lines that conform to the tested product. The certified product is also listed in the AAMA Certified Products Directory. When the design professional chooses to specify a certified product, the building owner has the added assurance that a prototype of the product has been tested and may opt to waive project specific mock-up tests. Before waiving project specific mock-up tests the design professional/owner should be confident that the certification test is a reasonable representation of the project conditions and must be aware of the additional risk. It is important for the building owner and specifier to understand that product unit certification does not guarantee building envelope performance by itself. Approved product listings such as those published by the Florida Building Product Approval Program, the Texas Department of Insurance, ICC-ES, and NFRC may also be helpful for project-specific product selection and should be consulted as appropriate.

The designer needs to understand that if he/she uses an approved product in the manner prescribed by the product or systems manufacturer, then no further testing or approvals are required to meet model building code requirements.

9.15 Accredited testing laboratories

Test results submitted by product manufacturers should be reported in writing by properly accredited testing organizations. ISO 17025 establishes the general requirements for the competence of testing and calibration laboratories. Accreditation bodies that utilize ISO 1705 are recommended for accreditation of testing organizations.

Testing and calibration laboratories that comply with this international standard will, therefore, also operate in accordance with ISO 9001. Each accredited laboratory must ensure the competence of its technical staff that performs specific tests and evaluates the test results. Personnel who perform specific tests shall be qualified on the basis of appropriate education, training, experience, and/or demonstrated skills.

9.16 References

American Architectural Manufacturers Association (AAMA) www.aamanet.org
1827 Walden Office Square, Suite 550
Schaumburg, IL 60173
AAMA 501.05, *Methods of Tests for Exterior Walls* (originally released by AAMA in 1968, updated in 1983, 1994, and 2005)
AAMA 501.1, *Standard Test Method for Exterior Windows, Curtain Walls and Doors for Water Penetration Using Dynamic Pressures*
AAMA 501.2-2003 and 501.3-1983, *Quality Assurance and Diagnostic Water Leakage Field Check of Installed Storefronts, Curtain Walls, and Sloped Glazing Systems*
AAMA 501.4-2000, *Recommended Static Test Method for Evaluating Curtain Wall and Storefront Systems Subjected to Seismic and Wind Induced Interstory Drifts*
AAMA 501.5-2005, *Test Method for Thermal Cycling of Exterior Walls*
AAMA 501.6-2001, *Recommended Dynamic Test Method for Determining the Seismic Drift Causing Glass Fallout from a Wall System Panel*
AAMA/WDMA/CSA 101/I.S.2/A440-2005, *Standard/Specification for Windows, Doors, and Unit Skylights*

ASTM International (ASTM) www.astm.org
100 Barr Harbor Drive
PO Box 6700
West Conshohocken, PA 19428-2959
ASTM E 283-04, *Standard Test Method for Determining Rate of Air Leakage Through Exterior Windows, Curtain Walls, and Doors under Specified Pressure Differences Across the Specimen*
ASTM E 330-02, *Standard Test Method for Structural Performance of Exterior Windows, Doors, Skylights and Curtain Walls by Uniform Static Air Pressure Difference*
ASTM E 331-00, *Standard Test Method for Water Penetration of Exterior Windows, Skylights, Doors, and Curtain Walls by Uniform Static Air Pressure Difference*
ASTM E 1886-05, *Standard Test Method for Performance of Exterior Windows, Curtain Walls, Doors, and Impact Protective Systems Impacted by Missile(s) and Exposed to Cyclic Pressure Differentials*
ASTM E 1996-08, *Standard Specification for Performance of Exterior Windows, Curtain Walls, Doors and Impact Protective Systems Impacted by Windborne Debris in Hurricanes*
ASTM E 2099-00, *Standard Practice for the Specification and Evaluation of Pre-Construction Laboratory Mockups of Exterior Wall Systems*
ASTM E 2178-03, *Standard Test Method for Air Permeance of Building Materials*

American Society of Civil Engineers (ASCE) www.asce.org
1801 Alexander Bell Drive
Reston, VA 20191-4400
ASCE/SEI 7-05, *Minimum Design Loads for Buildings and Other Structures*

FEMA (FEMA 4-90) www.fema.gov
500 C Street SW
Washington, DC 20472

International Code Council (ICC) www.iccsafe.org
5360 Workman Mill Road
Whittier, CA 90601

to resist windstorms in special buildings, 217–30
  buildings of special importance or with special functions, 218
  risk level, 223–4
  selected risk level for important facilities, 224
  site-specific design requirements, 225
  site survey, 224–5
  wind analysis, 218–23
  wind-borne debris impact requirements, 225
ground snow load, 67
gust effect factor, 157

'heat soaking' process, 181
heat strengthened glass, 35, 171, 179
high-velocity hurricane zone, 209–10
horizontal racking displacement, 20, 25
Hurricane Manufacturing Corporation, 229

ICC, 224
ICC-ES, 249
IG. see insulating glass
IG skylight, 178
impact protective system, 201
impact resistant covering, 208
impact resistant glazing, 208
impact risk analysis, 204
importance factor, 23, 156
insulating glass, 177, 185–6
Insulating Glass Certification Council, 102
interlayer shear modulus, 141, 144
internal pressure coefficient, 159–60
International Building Code, 1, 65, 101, 147, 149, 198, 207–8, 248, xii
International Residential Code, 65, 183, 207–8

laminated glass, 52
laminated glass/polycarbonate system, 227
large missile, 199–200
least squares line, 69
load–deflection curve, 243
load duration factors, 103, 105, 132, 134, 138, 141, 142, 144
Loma Prieta Earthquake, 53
long duration loads, 137

main wind force resisting system, 156

maximum considered earthquake, 4
mean recurrence interval, 65, 93, 151, 153
Methods of Tests for Exterior Walls, xii
Miami/Dade protocol, 196, 198, 208
Minimum Design Loads for Buildings and Other Structures, 65, 67, 101, 148, 193, 208–9, xii
Ministry of Land, Infrastructure and Transport, 74
missile propulsion device, 200
modulus of elasticity, 184, 185
Monte Carlo simulation, 153
mullion, 33–4

National Building Code, 248
National Building Code of Canada, 74
National Earthquake Hazard Reduction Program Recommended Provisions for Seismic Regulations for Buildings and Other Structures, 2–3
National Fire Protection Association, 26, 148
National Institute of Water and Atmospheric Research Ltd, 74
National Research Council of Canada, 74
National Weather Service, 70
NEHRP. see National Earthquake Hazard Reduction Program Recommended Provisions for Seismic Regulations for Buildings and Other Structures
NEHRP TS-8 committee, 14
neoprene, 52
Next-Generation Performance-Based Earthquake Engineering Guidelines, 26
Next-Generation Performance-Based Seismic Design Criterion, 41
NFPA 5000, 148
nickel sulfide, 181
Nisqually Earthquake, 41
nonstructural component seismic importance factor, 11–12
nonstructural elements, 49
Northridge Earthquake, 53

Occupancy Category, 8–10
  description, 9
Office of Code Compliance, 227

PA 201-94, 226
PA 202-94, 226